Mental Arithmetic Introductory Book

Answers

Schofield & Sims

CONTENTS

SECTION 1
Notes on Content	3
Tests 1 to 12	4
Diagnostic Chart for Section 1	16

SECTION 2
Notes on Content	17
Tests 1 to 12	18
Diagnostic Chart for Section 2	30

SECTION 3
Notes on Content	31
Tests 1 to 12	32
Diagnostic Chart for Section 3	44

JUST FACTS
Doubles – Near doubles – Near near doubles	45
Counting on 1 and related facts – Counting on 2 and related facts – Adding 10 and related facts	46
Making 10 and related facts – Adding 9 and related facts – Adding 8 and related facts	47

TEACHER'S NOTES

Introduction
Schofield & Sims **Mental Arithmetic** provides differentiated practice tests in key areas of maths, to be administered regularly. The series consists of seven pupil books – all of them conforming to a standard layout. This ensures that children are not presented with too many variables at once.

The **Mental Arithmetic Introductory Book** contains:

- 36 one-page tests, each comprising three parts – Parts A, B and C
- three Achievement Charts for recording children's results (see pupil book pages 16, 30 and 44)
- nine short Just Facts tests, which feature the addition and subtraction facts for number bonds 1 to 10.

Parts A, B and C
Each of the 36 tests mentioned above appears on a single page and is divided into three parts (A, B and C). Parts A and B use pictures, symbols and simple language wherever possible so that children with reading difficulties will not be disadvantaged. It is suggested that one test is taken each week and that Parts A, B and C are set on separate days. Since speed with accuracy is important, a time limit of 10 minutes per part is recommended.

Answering the test questions
The material in each section is graded so that, before any test question is attempted, the work will usually have been covered in class; for more details see the Notes on Content (pages 3, 17 and 31). The term 'mental arithmetic' implies that answers only are required; therefore the books are presented in a one-per-child format, so that answers can be written in the blanks. If the children are allowed spare paper for workings out, remember that their responses will be slower.

Please note: You should explain to the children that ▢ indicates a missing number.

Marking
A book of answers, like this one, is available to accompany each pupil book. When the children have completed a test, read out the answers as the children mark their own work, or give them the answer books to refer to themselves. Then ask the children to complete the Achievement Charts appearing at the end of each section in the pupil book – or you may yourself complete the Diagnostic Charts on pages 16, 30 and 44 of this book.

Mental Arithmetic Introductory Book Answers

The strategies covered in Section 1 encompass all the possible combinations of single-digit number bonds. (Operations involving zero have not been used.)

If the children learn to search for patterns and use strategies this will reduce the need to remember countless facts.

The tests in Section 1 cover the following topics.

TEST 1	Adding and subtracting 1	Count on 1/ Count back 1
TEST 2	Adding and subtracting 2	Count on 2/ Count back 2
TEST 3	Adding and subtracting 10	Count on/back by one whole ten (*not* ten ones)
TEST 4	Number bonds that make 10	Learn the bonds
TEST 5	Adding and subtracting 9	Add 10 then – 1 Subtract 10 then + 1
TEST 6	Adding and subtracting 8	Add 10 then – 2 Subtract 10 then + 2
TEST 7	Doubles	Learn the facts
TEST 8	Near doubles (a number bond where one number is 1 more than the other)	Use the double then + or – 1
TEST 9	Near near doubles (a number bond where one number is 2 more than the other)	Use the double then + or – 2 or Double the number between
TEST 10	Sets of 2	Multiplication
TEST 11	A mixture of question types from tests 1 to 9, to test progress	
TEST 12	A mixture of question types from tests 1 to 10, to test progress	

SECTION 1 | Notes on Content

Tests in Section 1

The tests in this section are designed to help children to use different strategies for remembering number bonds. It is hoped that these strategies will have been taught prior to the children attempting the tests.

The layout of the tests and the specific focus of each (see contents and the tables below) should help with the diagnosis of particular problems.

Part A
This section uses pictures as follows. All addition and subtraction is relevant to the strategy being tested.

Question	
1	addition of sets
2	subtraction of sets
3	missing number (using sets)
4	addition of money
5	subtraction of money
6	missing number (money)
7	naming shapes and fractions
8	measures language
9	days of the week
10	telling the time to the hour

Part B
This section uses patterns to help the child to extend knowledge of a single number bond to larger numbers.

Question	
1	
2	patterns to show how the addition strategy can be extended to larger numbers
3	
4	addition sequence relevant to the strategy
5	
6	patterns to show how the subtraction strategy can be extended to larger numbers
7	
8	subtraction sequence relevant to the strategy
9	missing number using +
10	missing number using –

Part C
This section uses addition and subtraction language with the appropriate set of number bonds. The order is kept the same between tests to make it easy to diagnose consistent mistakes.

Question	
1	add
2	take
3	total
4	difference
5	sum
6	more than
7	subtract
8	minus
9	less than
10	plus

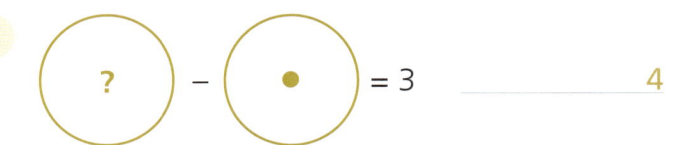

SECTION 1 | Test 1

Schofield & Sims

A   Answer

1. ⬤ (6 dots) + ⬤ (1 dot) = _____ 7

2. ⬤ (8 dots) − ⬤ (1 dot) = _____ 7

3. ⬤ (?) − ⬤ (1 dot) = 3 _____ 4

4. 5p + 1p = _____ 6p

5. 10p − 1p = _____ 9p

6. ? − 1p = 4p _____ 5p

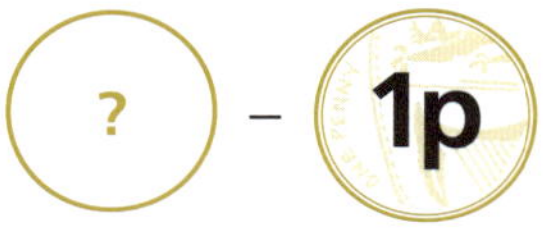

7. Write the name of this shape. _____ square

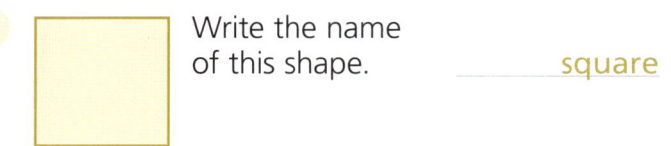

8. Which pencil is longer? _____ A
   - A (longer)
   - B (shorter)

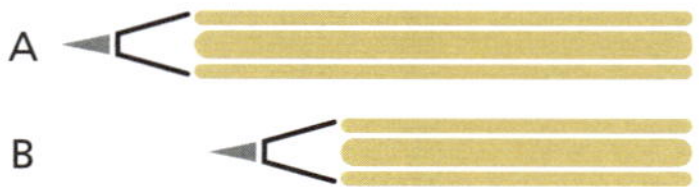

9. Which day comes after Tuesday? _____ Wednesday

10. What time is it? _____ 5 o'clock

B   Answer

1. 4 + 1 = _____ 5
2. 14 + 1 = _____ 15
3. 24 + 1 = _____ 25
4. 14, 15, 16, 17, ☐ _____ 18
5. 7 − 1 = _____ 6
6. 17 − 1 = _____ 16
7. 27 − 1 = _____ 26
8. 27, 26, 25, 24, ☐ _____ 23
9. ☐ + 1 = 4 _____ 3
10. ☐ − 1 = 8 _____ 9

C   Answer

1. Add 7 and 1. _____ 8
2. Take 1 from 5. _____ 4
3. The total of 8 and 1 is _____ 9
4. The difference between 1 and 6 is _____ 5
5. The sum of 1 and 9 is _____ 10
6. 3 is 1 more than _____ 2
7. Subtract 1 from 9. _____ 8
8. 4 minus 1 is _____ 3
9. 1 less than 8 is _____ 7
10. 1 plus 6 is _____ 7

4

Mental Arithmetic Introductory Book Answers

SECTION 1 | Test 2

A | Answer

1. ⬤ + ⬤ = __11__
2. ⬤ − ⬤ = __5__
3. ⬤ − ? = 4 __2__
4. 5p + 2p = __7__p
5. 10p − 2p = __8__p
6. ? − 2p = 8p __10__p
7. Write the name of this shape. __triangle__
8. Which ball is heavier? __A__
9. Which day comes before Saturday? __Friday__
10. What time is it? __10__ o'clock

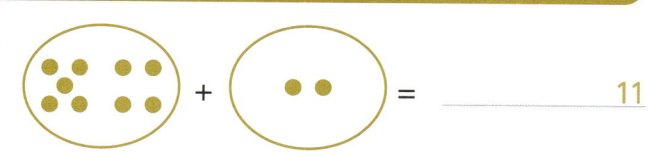

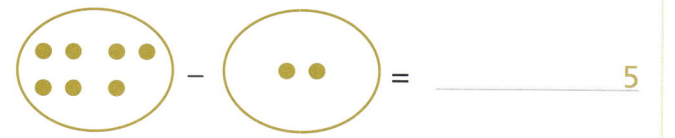

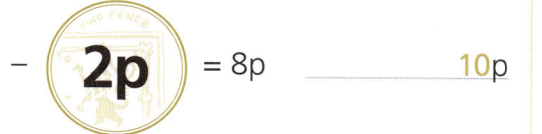

B | Answer

1. 6 + 2 = __8__
2. 16 + 2 = __18__
3. 26 + 2 = __28__
4. 16, 18, 20, 22, ▢ __24__
5. 5 − 2 = __3__
6. 15 − 2 = __13__
7. 25 − 2 = __23__
8. 25, 23, 21, 19, ▢ __17__
9. ▢ + 2 = 4 __2__
10. ▢ − 2 = 8 __10__

C | Answer

1. Add 8 and 2. __10__
2. Take 2 from 9. __7__
3. The total of 6 and 2 is __8__
4. The difference between 2 and 8 is __6__
5. The sum of 2 and 9 is __11__
6. 4 is 2 more than __2__
7. Subtract 2 from 5. __3__
8. 3 minus 2 is __1__
9. 2 less than 6 is __4__
10. 2 plus 7 is __9__

5

SECTION 1 | Test 3

Schofield & Sims

A | Answer

1. ⬤ + ⬤ = _____ 19
2. ⬤ − ⬤ = _____ 5
3. ⬤ − ? = 10 _____ 5
4. 5p + 10p = _____ 15p
5. 20p − 10p = _____ 10p
6. 2p + ? = 12p _____ 10p
7. Write the name of this shape. _____ rectangle/oblong
8. Which glass is full? _____ B 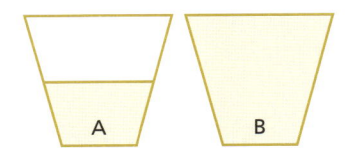
9. Which day comes after Monday? _____ Tuesday
10. What time is it? _____ 6 o'clock

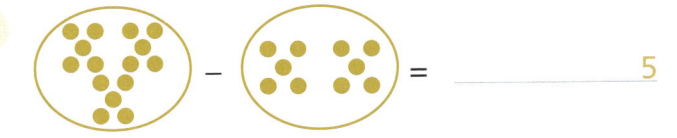

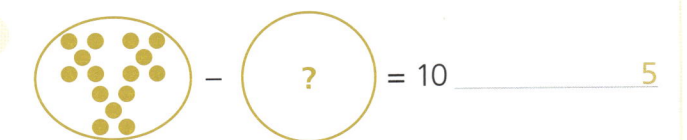

B | Answer

1. 3 + 10 = _____ 13
2. 13 + 10 = _____ 23
3. 23 + 10 = _____ 33
4. 13, 23, 33, 43, _____ 53
5. 11 − 10 = _____ 1
6. 21 − 10 = _____ 11
7. 31 − 10 = _____ 21
8. 51, 41, 31, 21, _____ 11
9. ___ + 10 = 46 _____ 36
10. ___ − 10 = 27 _____ 37

C | Answer

1. Add 10 and 3. _____ 13
2. Take 10 from 19. _____ 9
3. The total of 10 and 5 is _____ 15
4. The difference between 15 and 10 is _____ 5
5. The sum of 10 and 6 is _____ 16
6. 12 is 10 more than _____ 2
7. Subtract 10 from 18. _____ 8
8. 11 minus 10 is _____ 1
9. 10 is 4 less than _____ 14
10. 10 plus 7 is _____ 17

6

Mental Arithmetic Introductory Book Answers

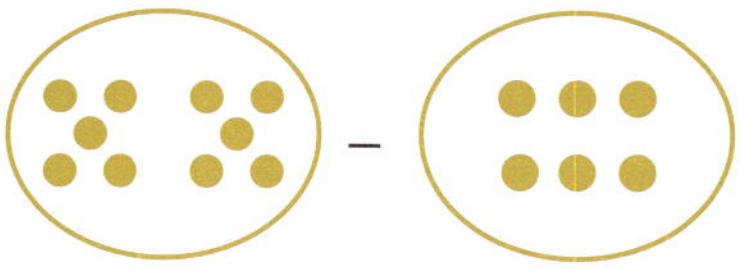

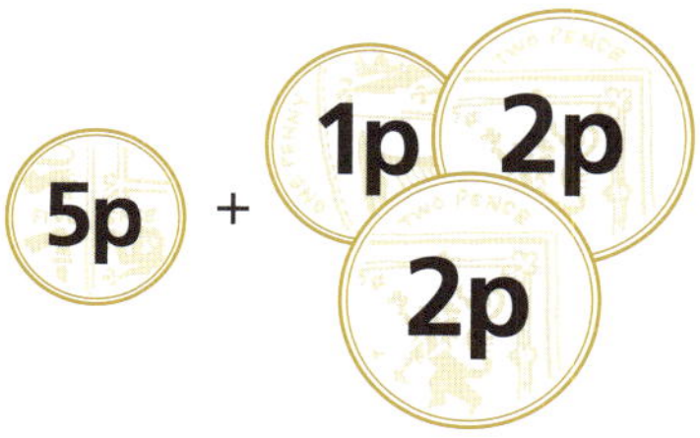

SECTION 1 | Test 4

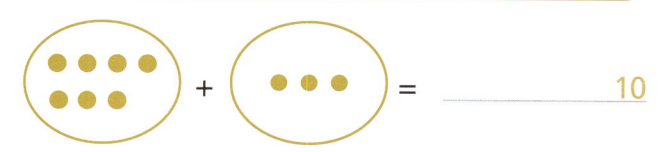

A

1. ●●●●● + ●●● = **10**
2. ●●●●●● − ●●●●● = **4**
3. ●●●●●● − ? = 2   **8**
4. 5p + (1p, 2p, 2p) = **10**p
5. 10p − 5p = **5**p
6. 10p − (?, ?) = 7p   **2p 1p**
7. Write the name of this shape. **circle**
8. Which pencil is shorter? **B**
9. Which day comes before Friday? **Thursday**
10. What time is it? **1 o'clock**

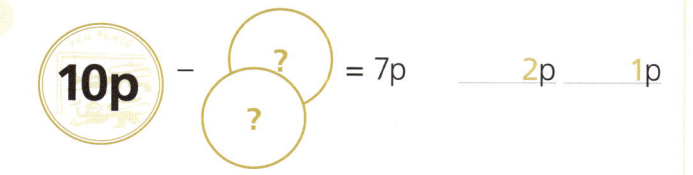

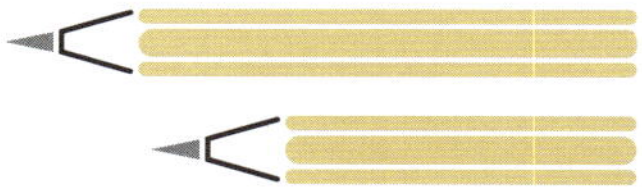

B

1. 6 + 4 =   **10**
2. 16 + 4 =   **20**
3. 46 + 4 =   **50**
4. 26, 36, 46, 56,   **66**
5. 10 − 3 =   **7**
6. 20 − 3 =   **17**
7. 30 − 3 =   **27**
8. 67, 57, 47, 37,   **27**
9. ▯ + 5 = 10   **5**
10. 10 − ▯ = 9   **1**

C

1. Add 3 and 7.   **10**
2. Take 9 from 10.   **1**
3. The total of 8 and 2 is   **10**
4. The difference between 5 and 10 is   **5**
5. The sum of 4 and 6 is   **10**
6. 10 is 3 more than   **7**
7. Subtract 8 from 10.   **2**
8. 10 minus 1 is   **9**
9. 6 is 4 less than   **10**
10. 2 plus 8 is   **10**

SECTION 1 | Test 5

Schofield & Sims

A | Answer

1. ⬤ + ⬤ = _____ 15
2. ⬤ − ⬤ = _____ 3
3. ? − ⬤ = 3 _____ 12
4. 5p + (5p 2p 2p) = _____ 14p
5. 20p − (5p 2p 2p) = _____ 11p
6. ? − (£5 £2 £2) = £1 £10
7. Write the name of this shape. _____ hexagon
8. Which ball is lighter? _____ A
9. Which day comes after Sunday? _____ Monday
10. What time is it? _____ 7 o'clock

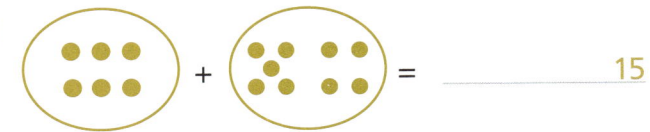

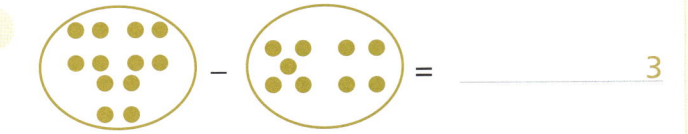

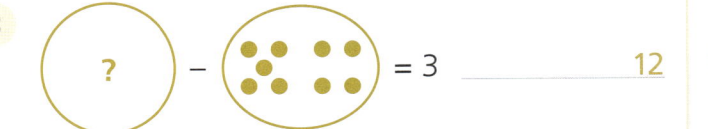

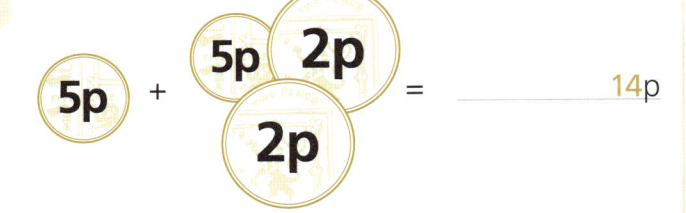

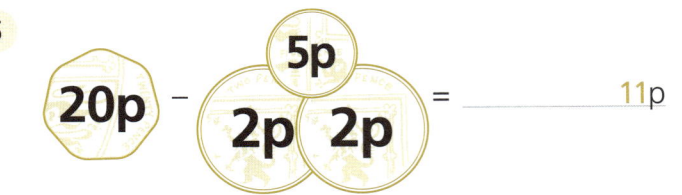

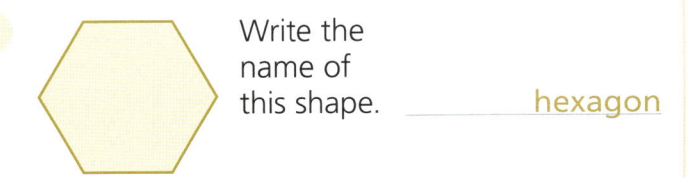

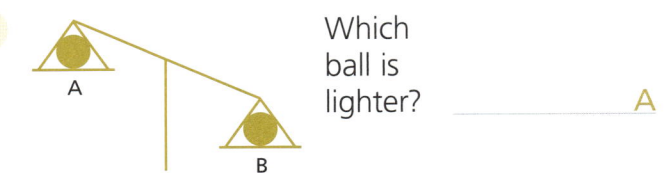

B | Answer

1. 7 + 9 = _____ 16
2. 17 + 9 = _____ 26
3. 27 + 9 = _____ 36
4. 9, 18, 27, 36, ☐ _____ 45
5. 14 − 9 = _____ 5
6. 24 − 9 = _____ 15
7. 44 − 9 = _____ 35
8. 54, 45, 36, 27, ☐ _____ 18
9. ☐ + 9 = 15 _____ 6
10. 12 − ☐ = 9 _____ 3

C | Answer

1. Add 4 and 9. _____ 13
2. Take 9 from 14. _____ 5
3. The total of 9 and 5 is _____ 14
4. The difference between 12 and 9 is _____ 3
5. The sum of 9 and 8 is _____ 17
6. 12 is 9 more than _____ 3
7. Subtract 9 from 13. _____ 4
8. 15 minus 9 is _____ 6
9. 9 is 7 less than _____ 16
10. 9 plus 3 is _____ 12

Mental Arithmetic Introductory Book Answers

SECTION 1 | Test 6

A — Answer

1. 14
2. 4
3. 11
4. 5p + (5p + 2p + 1p) = 13p
5. 20p − (5p + 2p + 1p) = 12p
6. ? − (2p + 2p + 2p + 2p) = 12p → 20p
7. Write the name of this shape. — pentagon
8. Which person is taller? — A
9. Which day comes before Thursday? — Wednesday
10. What time is it? — 3 o'clock

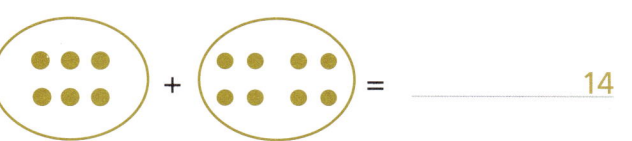

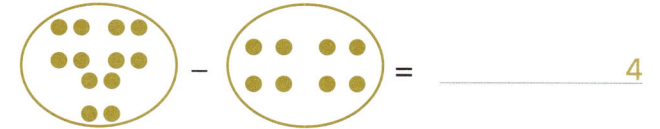

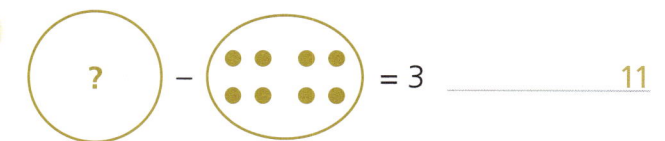

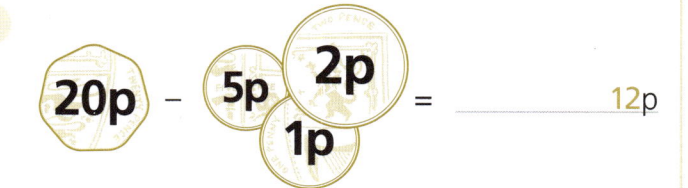

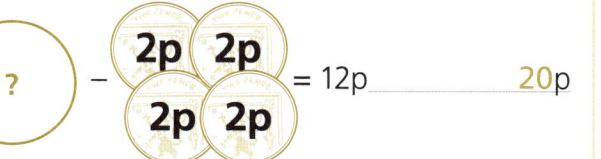

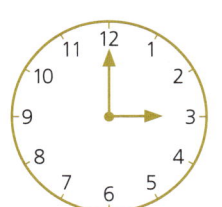

B — Answer

1. $5 + 8 =$ — 13
2. $15 + 8 =$ — 23
3. $35 + 8 =$ — 43
4. 8, 16, 24, 32, ▢ — 40
5. $14 − 8 =$ — 6
6. $24 − 8 =$ — 16
7. $44 − 8 =$ — 36
8. 48, 40, 32, 24, ▢ — 16
9. ▢ $+ 8 = 15$ — 7
10. $12 − ▢ = 8$ — 4

C — Answer

1. Add 7 and 8. — 15
2. Take 8 from 14. — 6
3. The total of 8 and 4 is — 12
4. The difference between 15 and 8 is — 7
5. The sum of 8 and 3 is — 11
6. 12 is 8 more than — 4
7. Subtract 8 from 13. — 5
8. 11 minus 8 is — 3
9. 8 is 5 less than — 13
10. 8 plus 6 is — 14

9

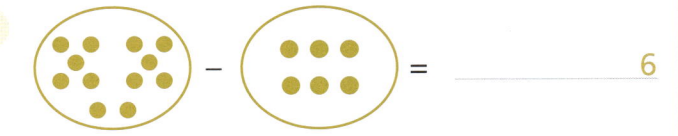

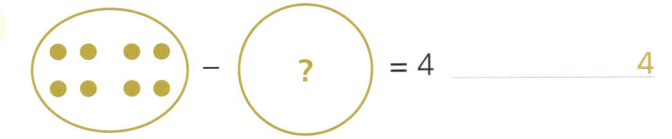

SECTION 1 | Test 7

A

#	Question	Answer
1	●●● + ●●● =	16
2	●●● − ●●● =	6
3	●●● − ? = 4	4
4	5p + 2p + 5p + 2p =	14p
5	£20 − £5 =	£10
6	? − 5p = 5p	10p
7	How much is shaded?	half
8	Which glass is half full?	B
9	If today is Monday, what was yesterday?	Sunday
10	What time is it?	4 o'clock

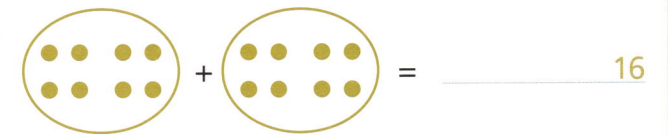

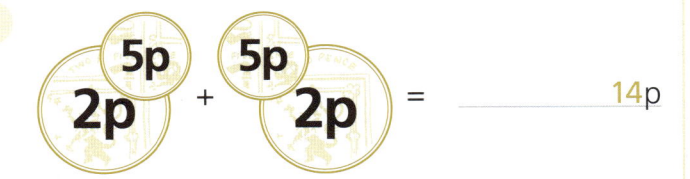

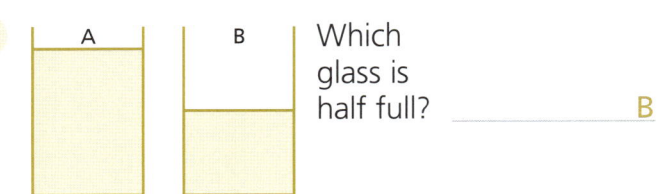

B

#	Question	Answer
1	6 + 6 =	12
2	16 + 6 =	22
3	26 + 6 =	32
4	1, 2, 4, 8, ☐	16
5	16 − 8 =	8
6	26 − 8 =	18
7	36 − 8 =	28
8	16, 8, 4, 2, ☐	1
9	☐ + 9 = 18	9
10	☐ − 7 = 7	14

C

#	Question	Answer
1	Add 8 and 8.	16
2	Take 7 from 14.	7
3	The total of 5 and 5 is	10
4	The difference between 18 and 9 is	9
5	The sum of 6 and 6 is	12
6	8 is 4 more than	4
7	Subtract 5 from 10.	5
8	6 minus 3 is	3
9	7 is 7 less than	14
10	9 plus 9 is	18

Mental Arithmetic Introductory Book Answers

SECTION 1 | Test 8

A | Answer

1. ● + ● = **15**
2. ● − ● = **8**
3. ● − ? = 4 **5**
4. 5p + 1p 5p = **11p**
5. 1p 10p 2p − 2p 5p = **6p**
6. 1p 10p − ? = 6p **5p**
7. How much is shaded? **quarter**
8. Which door is wider? **B**
9. If today is Wednesday, what was yesterday? **Tuesday**
10. What time is it? **11** o'clock

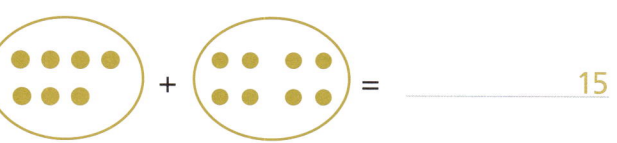

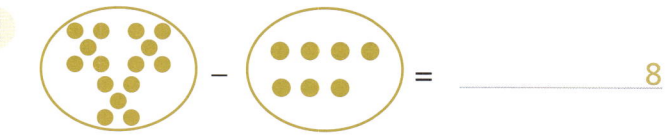

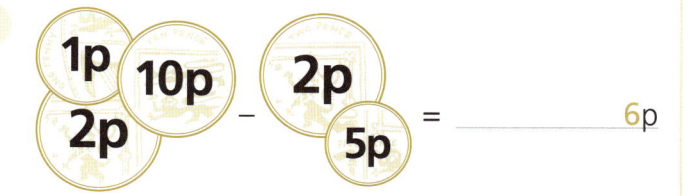

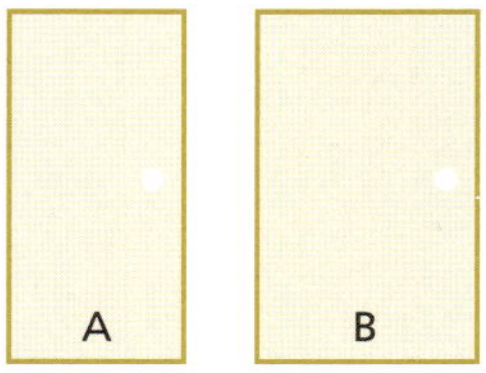

B | Answer

1. 3 + 4 = **7**
2. 13 + 4 = **17**
3. 53 + 4 = **57**
4. 3, 7, 11, 15, ▩ **19**
5. 9 − 4 = **5**
6. 19 − 4 = **15**
7. 39 − 4 = **35**
8. 29, 25, 21, 17, ▩ **13**
9. ▩ + 5 = 11 **6**
10. ▩ − 7 = 8 **15**

C | Answer

1. Add 6 and 7. **13**
2. Take 9 from 19. **10**
3. The total of 6 and 5 is **11**
4. The difference between 5 and 9 is **4**
5. The sum of 4 and 3 is **7**
6. 5 is 2 more than **3**
7. Subtract 7 from 15. **8**
8. 17 minus 8 is **9**
9. 5 is 6 less than **11**
10. 8 plus 7 is **15**

SECTION 1 | Test 9

Schofield & Sims

A — Answer

1. ⬤ + ⬤ = __12__
2. ⬤ − ⬤ = __6__
3. ⬤ − ? = 3   __5__
4. 5p + 5p + 2p = __12__p
5. 10p + 2p − 2p − 5p = __5__p
6. £10 − ? = £7   __£5__
7. How much is shaded?   __three-quarters__
8. How much longer is **A** than **B**?   __8__cm
   A ———— 14cm
   B —— 6cm
9. If today is Friday, what will tomorrow be?   __Saturday__
10. What time is it?   __9__ o'clock

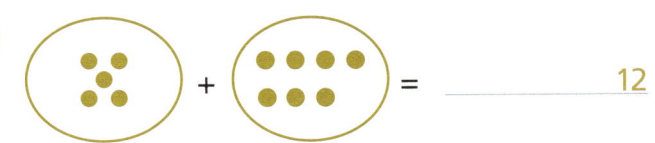

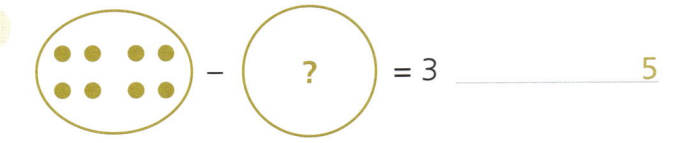

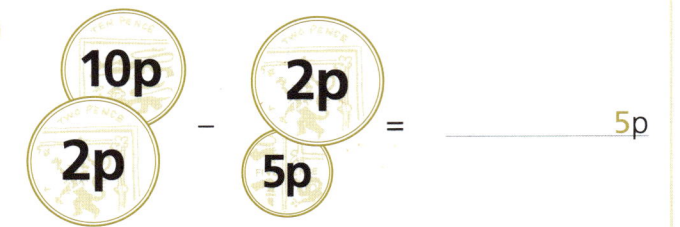

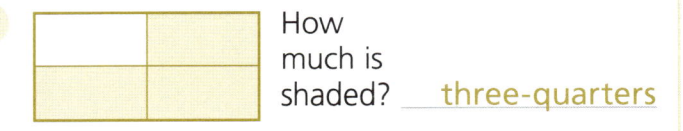

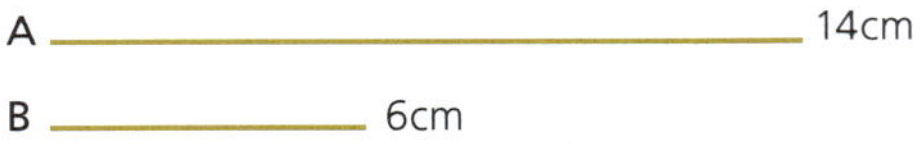

B — Answer

1. 5 + 3 = __8__
2. 15 + 3 = __18__
3. 35 + 3 = __38__
4. 5, 8, 11, 14, __17__
5. 7 − 5 = __2__
6. 17 − 5 = __12__
7. 57 − 5 = __52__
8. 37, 32, 27, 22, __17__
9. ▢ + 5 = 8   __3__
10. ▢ − 7 = 5   __12__

C — Answer

1. Add 6 and 8.   __14__
2. Take 9 from 15.   __6__
3. The total of 3 and 5 is   __8__
4. The difference between 5 and 12 is   __7__
5. The sum of 4 and 6 is   __10__
6. 6 is 2 more than   __4__
7. Subtract 7 from 12.   __5__
8. 18 minus 8 is   __10__
9. 1 is 3 less than   __4__
10. 9 plus 7 is   __16__

Mental Arithmetic Introductory Book Answers

SECTION 1 | Test 10

A

1. Turn the picture into a sum. **4** sets of **2**
2. Turn the picture into a sum. **2** sets of **3** 
3.  ●● × 6 =   **12**
4. ●●●●●●●● ÷ 2 =   **4**
5. How much altogether?   **£20** 
6. 2p × ☐ = 10p   **5**
7. 20p ÷ 2 =   **10**p
8. B is twice as heavy as A. How heavy is B? (A = 10g)   **20g**
9. If today is Sunday, what was yesterday?   **Saturday**
10. What time is it?   **12 o'clock**

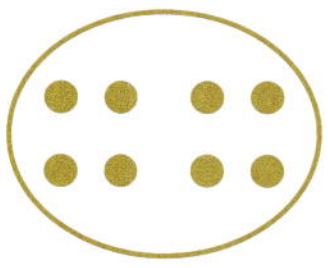

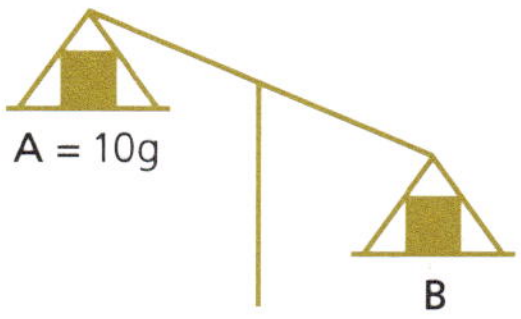

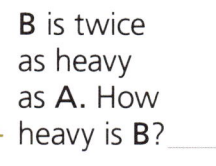

B

1. 2, 4, 6, 8, ☐   **10**
2. 2 + 2 + 2 =   **6**
3. 3 × 2 =   **6**
4. 2 + 2 + 2 + 2 + 2 + 2 + 2 =   **14**
5. 7 × 2 =   **14**
6. 20, 18, 16, ☐, 12   **14**
7. 6 − 2 − 2 − 2 =   **0**
8. 6 ÷ 2 =   **3**
9. 14 ÷ 2 =   **7**
10. 12 ÷ 2 =   **6**

C

1. 8 sets of 2 are   **16**
2. How many twos in 10?   **5**
3. 4 twos are   **8**
4. How many 2p coins make 20p?   **10**
5. 9 times 2 is   **18**
6. 6 multiplied by 2 is   **12**
7. 8 shared among 2 is   **4**
8. Half of 18 is   **9**
9. 6 divided by 2 is   **3**
10. How many socks are there in 10 pairs?   **20**

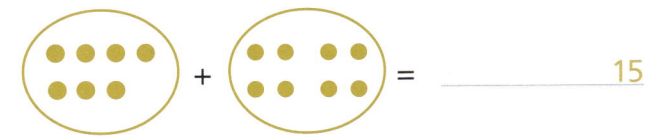

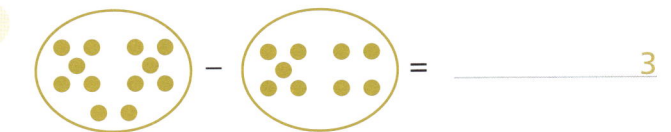

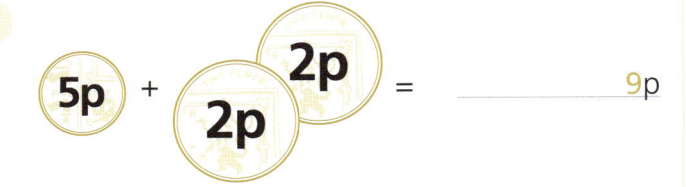

SECTION 1 | Test 11

A

#	Question	Answer
1	(7 dots) + (8 dots) =	15
2	(10 dots) − (7 dots) =	3
3	(10 dots) − ? = 6	4
4	5p + 2p + 2p =	9p
5	20p − (1p + 1p + 5p) =	13p
6	(1p + 1p + 10p) − ? = 7p	5p
7	Which shape is cut in half?	B
8	A = 5g, B = 8g. How much heavier is B?	3g
9	If today is Thursday, what will tomorrow be?	Friday
10	What time is it?	8 o'clock

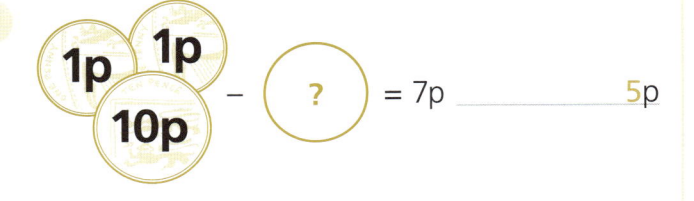

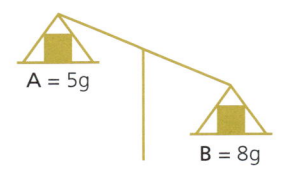

B

#	Question	Answer
1	6 + 7 =	13
2	46 + 7 =	53
3	3 + 9 =	12
4	73 + 9 =	82
5	1, 3, 5, 7, ▢	9
6	18 − 9 =	9
7	38 − 9 =	29
8	17 − 10 =	7
9	57 − 10 =	47
10	19, 17, 15, 13, ▢	11

C

#	Question	Answer
1	Add 5 and 8.	13
2	Take 6 from 11.	5
3	The total of 6 and 8 is	14
4	The difference between 13 and 4 is	9
5	The sum of 7 and 5 is	12
6	16 is 9 more than	7
7	Subtract 3 from 7.	4
8	17 minus 9 is	8
9	5 is 3 less than	8
10	7 plus 7 is	14

Mental Arithmetic Introductory Book Answers

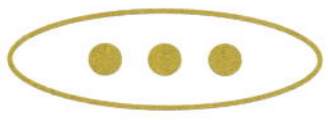

SECTION 1 | Test 12

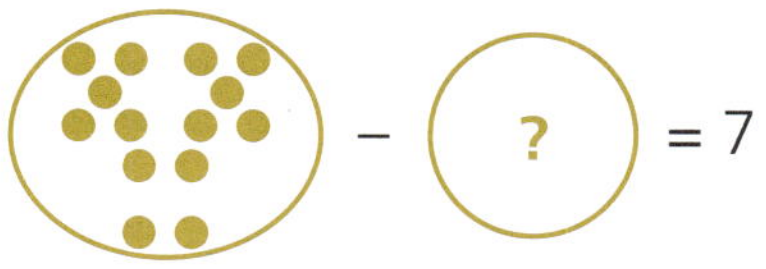

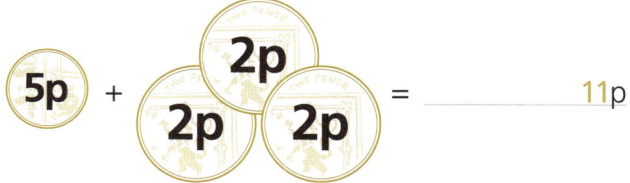

A

1. (●●●) × 2 =   6
2. (●●●●●●●●●●) ÷ 2 =   5
3. (●●●●●●●●●●●●) − ? = 7   7
4. 5p + 2p + 2p + 2p =   11p
5. £20 + £10 − £5 = £5
6. 2p 2p 2p 2p 2p − ? = 5p   5p
7. Which shape is cut into quarters?   C
8. A is twice as long as B. How long is A?   8cm
9. If today is Saturday, what will tomorrow be?   Sunday
10. What time is it?   2 o'clock

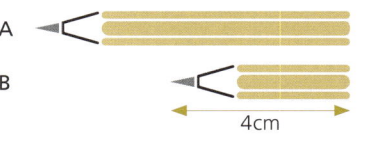

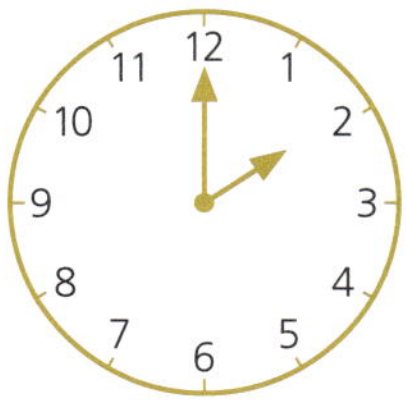

B

1. 8 + 4 =   12
2. 38 + 4 =   42
3. 9 − 6 =   3
4. 29 − 6 =   23
5. ☐ + 4 = 10   6
6. ☐ − 9 = 8   17
7. 4, 6, 8, 10, ☐   12
8. 6 × 2 =   12
9. 16, 14, 12, 10, ☐   8
10. 18 ÷ 2 =   9

C

1. 9 sets of 2 are   18
2. How many twos in 14?   7
3. The total of 5 and 9 is   14
4. The difference between 6 and 15 is   9
5. The sum of 4 and 8 is   12
6. 9 is 4 more than   5
7. Subtract 7 from 16.   9
8. 12 shared among 2 is   6
9. 9 is 4 less than   13
10. 8 multiplied by 2 is   16

15

Name of child _____		Diagnostic Chart for Section 1 Indicate where child has difficulty									
		1	2	3	4	5	6	7	8	9	10
Test 1	Part A										
	Part B										
	Part C										
Test 2	Part A										
	Part B										
	Part C										
Test 3	Part A										
	Part B										
	Part C										
Test 4	Part A										
	Part B										
	Part C										
Test 5	Part A										
	Part B										
	Part C										
Test 6	Part A										
	Part B										
	Part C										
Test 7	Part A										
	Part B										
	Part C										
Test 8	Part A										
	Part B										
	Part C										
Test 9	Part A										
	Part B										
	Part C										
Test 10	Part A										
	Part B										
	Part C										
Test 11	Part A										
	Part B										
	Part C										
Test 12	Part A										
	Part B										
	Part C										

From: **Mental Arithmetic Introductory Book Answers**. Copyright © Schofield & Sims Ltd, 2016. This page may be photocopied after purchase.

Mental Arithmetic Introductory Book Answers

These tests enable children to use mental methods to add and subtract 'teen' numbers. They should not become bogged down in exchanging tens and so on, but should use their knowledge of number bonds to find the answer.

The tests in Section 2 cover the following topics.

TEST 1	Addition and subtraction, making 11	Add/subtract the 10 then the units
TEST 2	Addition and subtraction, making 12	or
TEST 3	Addition and subtraction, making 13	Add/subtract the units then the 10
TEST 4	Addition and subtraction, making 14	Use the number bond strategies previously learnt
TEST 5	Addition and subtraction, making 15	
TEST 6	Addition and subtraction, making 16	
TEST 7	Addition and subtraction, making 17	
TEST 8	Addition and subtraction, making 18	Add/subtract 2 tens then + or – 2
TEST 9	Addition and subtraction, making 19	Add/subtract 2 tens then + or – 1
TEST 10	Sets of 10	Multiplication
TEST 11	A mixture of question types from tests 1 to 9, to test progress	
TEST 12	A mixture of question types from tests 1 to 10, to test progress	

SECTION 2 | Notes on Content

Tests in Section 2

The tests in this section are designed to help children to develop the ability to add and subtract 'teen' numbers and to have an awareness of how to use place value. It is hoped that these strategies will have been taught previously.

The layout of the tests and the specific focus of each (see contents and the tables below) should help with the diagnosis of particular problems.

Part A	
This section uses pictures as follows. Numerical examples relate to the 'teen' number being tested.	
Question	
1	identifying the number
2	knowing the name of the number
3	ability to show the number on an abacus
4	totalling money
5	making up to 20p
6	shopping language — subtraction of 5p and 10p
7	shopping language — subtraction of 9p
8	shopping language — addition of 5p and 10p
9	months of the year
10	telling the time to the half hour

Part B	
This section uses patterns to help the child to extend knowledge of a single number bond to larger numbers.	
Question	
1	developing a strategy for addition
2	developing a strategy for addition
3	developing a strategy for addition
4	developing a strategy for addition
5	using the addition strategy
6	developing a strategy for subtraction
7	developing a strategy for subtraction
8	developing a strategy for subtraction
9	missing number using –
10	missing number using +

Part C	
This section uses addition and subtraction language and revises making the 'teen' number as well as addition of the 'teen' number. The order is kept the same between tests to make it easy to diagnose consistent mistakes.	
Question	
1	add
2	take
3	total
4	difference
5	sum
6	more than
7	subtract
8	minus
9	less than
10	plus

SECTION 2 | Test 1

Schofield & Sims

A | Answer

1. How many sticks? — 11
2. Write the number of cubes in words. — eleven
3. Draw beads on the abacus to show the number eleven. 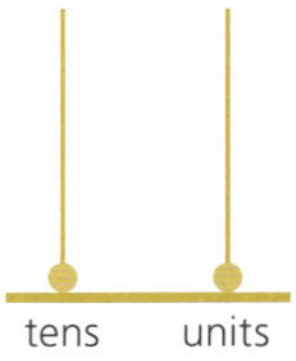
4. The total value of the money is £11. What is missing? — £5 
5. How much more do you need to make 20p? — 9p
6. You have 10p, 1p. You spend 5p. How much have you got left? — 6p
7. You have 11p. You buy (9p). How much change do you get? — 2p
8. You have 5p, 1p, 5p. You win 5p. How much do you have now? — 16p
9. Which month comes after March? — April
10. What time is it? — half past five

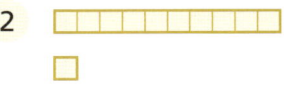

B | Answer

1. 6 + 10 + 1 = — 17
2. 16 + 10 + 1 = — 27
3. 26 + 10 + 1 = — 37
4. 36 + 11 = — 47
5. 24 + 11 = — 35
6. 17 − 10 − 1 = — 6
7. 27 − 10 − 1 = — 16
8. 37 − 10 − 1 = — 26
9. ☐ − 11 = 46 — 57
10. ☐ + 11 = 45 — 34

C | Answer

1. Add 7 and 11. — 18
2. Take 5 from 11. — 6
3. The total of 8 and 11 is — 19
4. The difference between 11 and 6 is — 5
5. The sum of 11 and 9 is — 20
6. 11 is 7 more than — 4
7. Subtract 9 from 11. — 2
8. 11 minus 4 is — 7
9. 8 less than 11 is — 3
10. 11 plus 6 is — 17

18

Mental Arithmetic Introductory Book Answers

SECTION 2 | Test 2

A

1.  How many sticks? — 12
2. 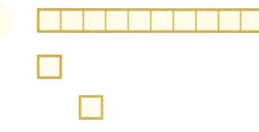 Write the number of cubes in words. — twelve
3. Draw beads on the abacus to show the number twelve. 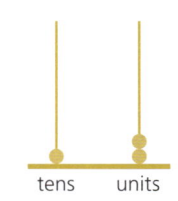
4. The total value of the coins is 12p. Which coins are missing? (5p, 5p, ?) — 2p 
5.  (10p, 2p) How much more do you need to make 20p? — 8p
6. You have  12p. You spend  5p. How much have you got left? — 7p
7. You have  5p, 2p, 5p. You buy (9p). How much change do you get? — 3p
8. You have  5p, 2p, 5p. You win  5p. How much do you have now? — 17p
9. Which month comes before June? — May
10.  What time is it? — half past ten

B

1. 5 + 10 + 2 = — 17
2. 15 + 10 + 2 = — 27
3. 25 + 10 + 2 = — 37
4. 35 + 12 = — 47
5. 26 + 12 = — 38
6. 19 − 10 − 2 = — 7
7. 29 − 10 − 2 = — 17
8. 39 − 12 = — 27
9. ☐ − 12 = 49 — 61
10. ☐ + 12 = 47 — 35

C

1. Add 14 and 12. — 26
2. Take 7 from 12. — 5
3. The total of 9 and 12 is — 21
4. The difference between 12 and 8 is — 4
5. The sum of 12 and 6 is — 18
6. 12 is 4 more than — 8
7. Subtract 3 from 12. — 9
8. 12 minus 5 is — 7
9. 10 less than 12 is — 2
10. 12 plus 7 is — 19

SECTION 2 | Test 3

Schofield & Sims

A | Answer

1. How many sticks? **13**

2. Write the number of cubes in words. **thirteen**

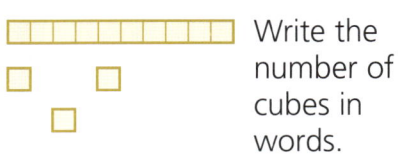

3. Draw the beads on the abacus to show the number thirteen.

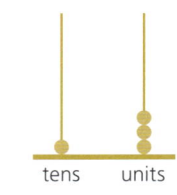

4. The total value of the money is £13. What is missing? £**1**

5. How much more do you need to make 20p? **8**p

6. You have 10p 2p 1p. You spend 5p. How much have you got left? **8**p

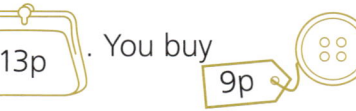

7. You have 13p. You buy 9p. How much change do you get? **4**p

8. You have 10p 2p 1p. You are given  5p. How much do you have now? **18**p

9. Which month comes before January? **December**

10. What time is it? **half past three**

B | Answer

1. 4 + 10 + 3 = **17**
2. 14 + 10 + 3 = **27**
3. 24 + 13 = **37**
4. 34 + 13 = **47**
5. 25 + 13 = **38**
6. 17 − 10 − 3 = **4**
7. 27 − 10 − 3 = **14**
8. 37 − 13 = **24**
9. ☐ − 13 = 43 **56**
10. ☐ + 13 = 47 **34**

C | Answer

1. Add 6 and 13. **19**
2. Take 4 from 13. **9**
3. The total of 5 and 13 is **18**
4. The difference between 13 and 8 is **5**
5. The sum of 13 and 7 is **20**
6. 13 is 7 more than **6**
7. Subtract 4 from 13. **9**
8. 13 minus 3 is **10**
9. 9 less than 13 is **4**
10. 13 plus 10 is **23**

Mental Arithmetic Introductory Book Answers

SECTION 2 | Test 4

A

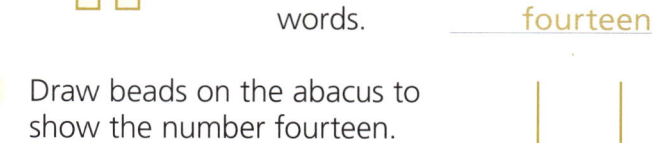

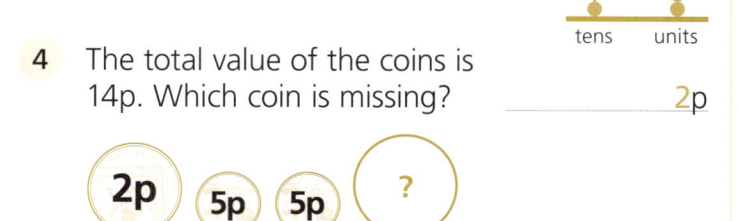

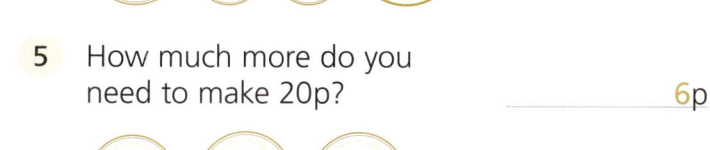

1. How many sticks? **14**
2. Write the number of cubes in words. **fourteen**
3. Draw beads on the abacus to show the number fourteen.
4. The total value of the coins is 14p. Which coin is missing? **2p**
5. How much more do you need to make 20p? **6p** 
6. You have 2p, 10p, 2p. You lose 5p. How much have you got left? **9p**
7. You have  14p. You buy (9p). How much change do you get? **5p**
8. You have  2p, 10p, 2p. You are given  5p. How much do you have now? **19p**
9. Which month comes after September? **October**
10. What time is it? **half past eleven**

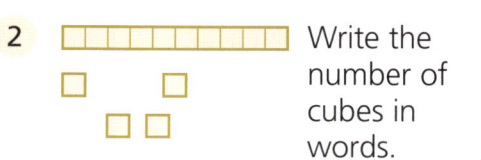

B

1. 6 + 10 + 4 = **20**
2. 16 + 14 = **30**
3. 26 + 14 = **40**
4. 36 + 14 = **50**
5. 56 + 14 = **70**
6. 19 – 10 – 4 = **5**
7. 29 – 10 – 4 = **15**
8. 39 – 14 = **25**
9. ☐ – 14 = 45 **59**
10. ☐ + 14 = 45 **31**

C

1. Add 5 and 14. **19**
2. Take 5 from 14. **9**
3. The total of 6 and 14 is **20**
4. The difference between 14 and 6 is **8**
5. The sum of 14 and 9 is **23**
6. 14 is 7 more than **7**
7. Subtract 9 from 14. **5**
8. 14 minus 4 is **10**
9. 8 less than 14 is **6**
10. 14 plus 3 is **17**

SECTION 2 | Test 5

Schofield & Sims

A | Answer

1. How many sticks? **15**

2. 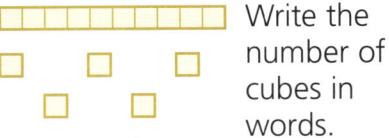 Write the number of cubes in words. **fifteen**

3. Draw beads on the abacus to show the number fifteen.

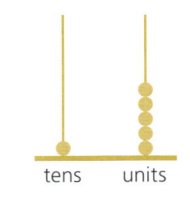

4. The total value of the money is £15. What is missing? £1, £2, ?, £2 — **£10**

5. 10p, 5p. How much more do you need to make 20p? **5p**

6. You have 10p, 5p. You lose 5p. How much have you got left? **10p**

7. You have  15p. You buy  9p. How much change do you get? **6p**

8. You have 10p, 5p. You are given 5p. How much do you have now? **20p**

9. Which month comes after July? **August**

10. 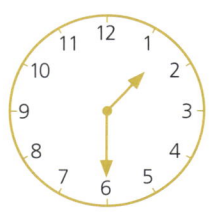 What time is it? **half past one**

B | Answer

1. 5 + 10 + 5 = **20**
2. 15 + 15 = **30**
3. 25 + 15 = **40**
4. 35 + 15 = **50**
5. 55 + 15 = **70**
6. 15 − 10 − 5 = **0**
7. 25 − 15 = **10**
8. 35 − 15 = **20**
9. ☐ − 15 = 45 **60**
10. ☐ + 15 = 60 **45**

C | Answer

1. Add 10 and 15. **25**
2. Take 7 from 15. **8**
3. The total of 3 and 15 is **18**
4. The difference between 15 and 6 is **9**
5. The sum of 15 and 9 is **24**
6. 15 is 5 more than **10**
7. Subtract 9 from 15. **6**
8. 15 minus 8 is **7**
9. 4 less than 15 is **11**
10. 15 plus 6 is **21**

22

Mental Arithmetic Introductory Book Answers

SECTION 2 | Test 6

A

		Answer
1	How many sticks?	16
2	Write the number of cubes in words.	sixteen
3	Draw beads on the abacus to show the number sixteen.	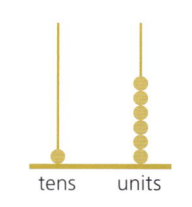
4	The total value of the coins is 16p. Which coin is missing?	5p
5	How much more do you need to make 20p?	4p
6	You have 10p, 5p, 1p. You lose 5p. How much have you got left?	11p
7	You have 16p. You buy 9p. How much change do you get?	7p
8	You have 10p, 5p, 1p. You are given 5p. How much do you have now?	21p
9	Which month comes before October?	September
10	What time is it?	half past eight

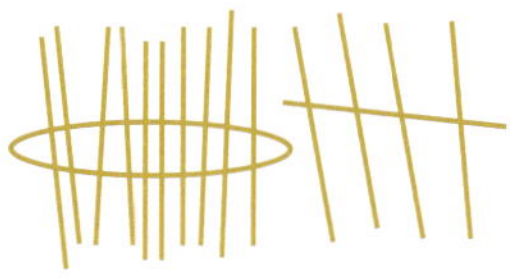

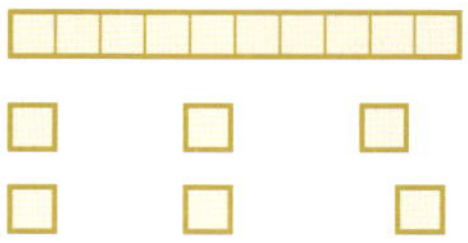

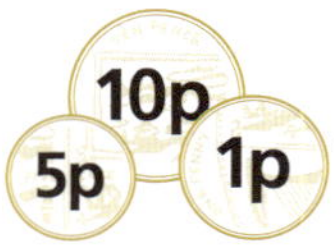

B

		Answer
1	2 + 10 + 6 =	18
2	12 + 16 =	28
3	22 + 16 =	38
4	32 + 16 =	48
5	52 + 16 =	68
6	18 − 10 − 6 =	2
7	28 − 16 =	12
8	38 − 16 =	22
9	☐ − 16 = 42	58
10	☐ + 16 = 68	52

C

		Answer
1	Add 10 and 16.	26
2	Take 7 from 16.	9
3	The total of 3 and 16 is	19
4	The difference between 16 and 6 is	10
5	The sum of 16 and 9 is	25
6	16 is 5 more than	11
7	Subtract 9 from 16.	7
8	16 minus 8 is	8
9	4 less than 16 is	12
10	16 plus 6 is	22

SECTION 2 | Test 7

A

#	Question	Answer
1	How many sticks?	17
2	Write the number of cubes in words.	seventeen
3	Draw beads on the abacus to show the number seventeen.	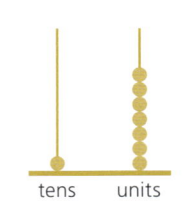
4	The total value of the coins is 17p. Which coin is missing?	5p
5	How much more do you need to make 20p?	3p
6	You have 10p, 5p, 2p. You lose 5p. How much have you got left?	12p
7	You have 17p. You buy apple 9p. How much change do you get?	8p
8	You have 10p, 5p, 2p. You are given 5p. How much do you have now?	22p
9	Which month comes before April?	March
10	What time is it?	half past four

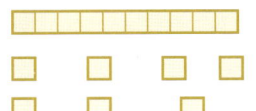

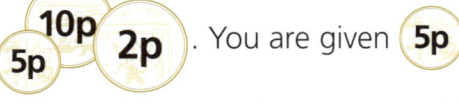

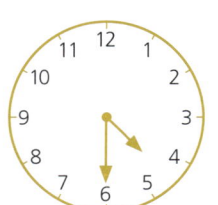

B

#	Question	Answer
1	4 + 10 + 7 =	21
2	14 + 17 =	31
3	24 + 17 =	41
4	34 + 17 =	51
5	54 + 17 =	71
6	20 − 10 − 7 =	3
7	30 − 17 =	13
8	40 − 17 =	23
9	▢ − 17 = 43	60
10	▢ + 17 = 61	44

C

#	Question	Answer
1	Add 7 and 17.	24
2	Take 5 from 17.	12
3	The total of 10 and 17 is	27
4	The difference between 17 and 6 is	11
5	The sum of 17 and 9 is	26
6	17 is 7 more than	10
7	Subtract 9 from 17.	8
8	17 minus 4 is	13
9	8 less than 17 is	9
10	17 plus 6 is	23

Mental Arithmetic Introductory Book Answers

SECTION 2 | Test 8

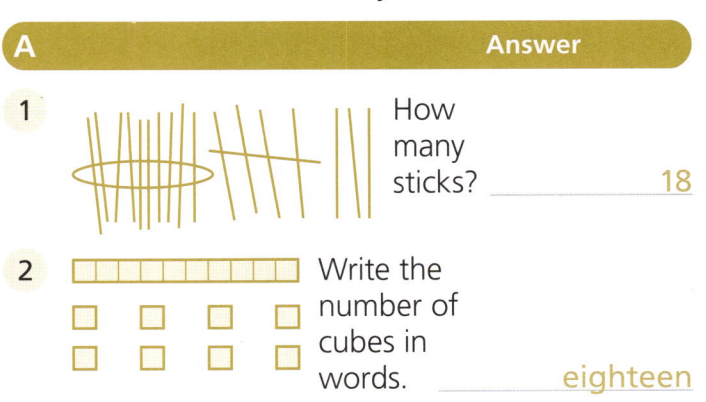

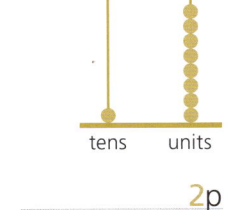

A | Answer
1. How many sticks? **18**
2. Write the number of cubes in words. **eighteen**
3. Draw beads on the abacus to show the number eighteen.
4. The total value of the coins is 18p. Which coin is missing? **2p** 
5. How much more do you need to make 20p? **2p** 
6. You have  10p, 2p, 5p, 1p. You spend  5p. How much have you got left? **13p**
7. You have  18p. You buy 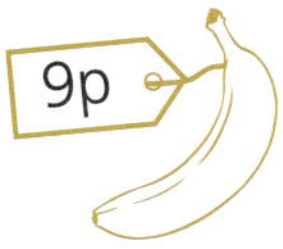 9p. How much change do you get? **9p**
8. You have 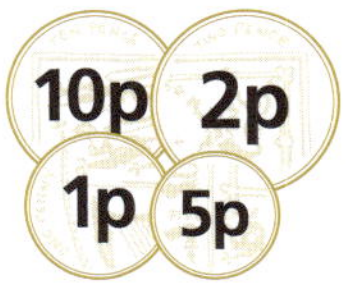 10p, 2p, 1p, 5p. You win  5p. How much do you have now? **23p**
9. Which month comes after November? **December**
10. What time is it? **half past nine**

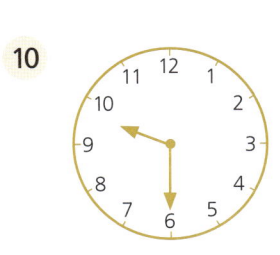

B | Answer
1. 4 + 10 + 8 = **22**
2. 4 + 10 + 10 − 2 = **22**
3. 24 + 10 + 8 = **42**
4. 24 + 10 + 10 − 2 = **42**
5. 24 + 18 = **42**
6. 20 − 10 − 8 = **2**
7. 20 − 10 − 10 + 2 = **2**
8. 20 − 18 = **2**
9. ☐ − 18 = 42   **60**
10. ☐ + 18 = 42   **24**

C | Answer
1. Add 10 and 18. **28**
2. Take 8 from 18. **10**
3. The total of 8 and 18 is **26**
4. The difference between 18 and 9 is **9**
5. The sum of 18 and 4 is **22**
6. 18 is 7 more than **11**
7. Subtract 6 from 18. **12**
8. 18 minus 5 is **13**
9. 3 less than 18 is **15**
10. 18 plus 2 is **20**

25

SECTION 2 | Test 9

A

		Answer
1	How many sticks?	19
2	Write the number of cubes in words.	nineteen
3	Draw beads on the abacus to show the number nineteen.	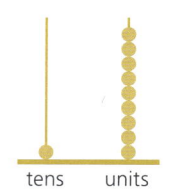
4	The total value of the coins is 19p. Which coin is missing? (10p, 5p, 2p, ?)	2p
5	How much more do you need to make 20p? (2p, 2p, 5p, 5p, 5p)	1p
6	You have 2p, 10p, 2p, 5p. You spend 10p. How much have you got left?	9p
7	You have 19p. You buy 9p button. How much change do you get?	10p
8	You have 10p, 5p, 2p, 2p. You win 10p. How much do you have now?	29p
9	Which month comes after May?	June
10	What time is it?	half past seven

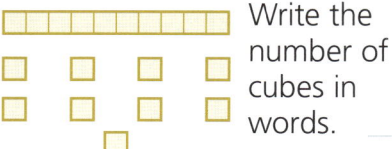

B

		Answer
1	3 + 10 + 9 =	22
2	3 + 10 + 10 − 1 =	22
3	23 + 10 + 9 =	42
4	23 + 10 + 10 − 1 =	42
5	23 + 19 =	42
6	40 − 10 − 9 =	21
7	40 − 10 − 10 + 1 =	21
8	40 − 19 =	21
9	☐ − 19 = 41	60
10	☐ + 19 = 41	22

C

		Answer
1	Add 5 and 19.	24
2	Take 10 from 19.	9
3	The total of 4 and 19 is	23
4	The difference between 19 and 6 is	13
5	The sum of 19 and 9 is	28
6	19 is 8 more than	11
7	Subtract 6 from 19.	13
8	19 minus 4 is	15
9	5 less than 19 is	14
10	19 plus 3 is	22

Mental Arithmetic Introductory Book Answers

SECTION 2 | Test 10

A

1. Turn the picture into a sum.   **3** sets of **10**

2. Turn the picture into a sum.   **10** sets of **3**

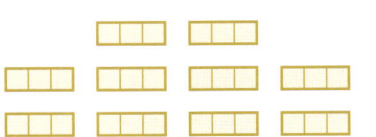

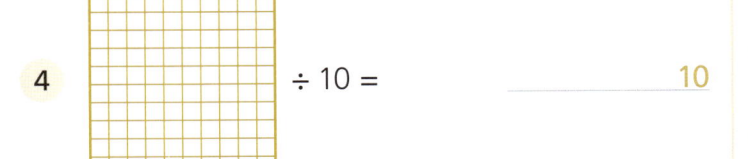

3. ⬭ × 10 =   **60**

4. ▦ ÷ 10 =   **10**

5. How much altogether?   **40**p

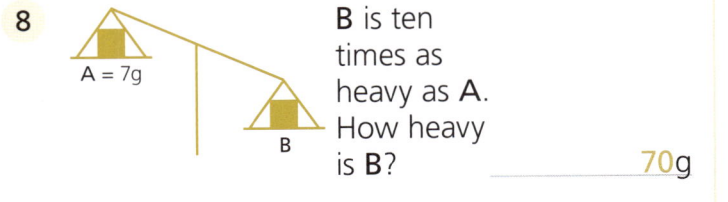

6. 10p × __ = 70p   **7**

7. £1 ÷ 10 =   **10**p

8. B is ten times as heavy as A. How heavy is B? (A = 7g)   **70**g

9. Which month comes after December?   **January**

10. What time is it?   **half past twelve**

B

1. 10, 20, 30, 40, __   **50**
2. 10 + 10 + 10 =   **30**
3. 3 × 10 =   **30**
4. 10 + 10 + 10 + 10 + 10 =   **50**
5. 5 × 10 =   **50**
6. 90, 80, 70, 60, __   **50**
7. 30 − 10 − 10 − 10 =   **0**
8. 30 ÷ 10 =   **3**
9. 70 ÷ 10 =   **7**
10. 90 ÷ 10 =   **9**

C

1. 8 sets of 10 are   **80**
2. How many tens in 70?   **7**
3. 4 tens are   **40**
4. How many 10p coins make 50p?   **5**
5. 9 times 10 is   **90**
6. 6 multiplied by 10 is   **60**
7. 80 shared among 10 is   **8**
8. How many 10p coins have the same value as £1?   **10**
9. 60 divided by 10 is   **6**
10. How many wheels are there altogether on 10 bicycles?   **20**

SECTION 2 | Test 11

A

#	Question	Answer
1	How many sticks?	15
2	Write the number of cubes in words.	nineteen
3	Draw beads on the abacus to show the number seventeen.	
4	The total value of the coins is 12p. Which coin is missing?	5p
5	How much more do you need to make 20p?	4p
6	You have 2p, 2p, 5p, 5p. You lose 10p. How much have you got left?	4p
7	You have 18p. You buy 7p cupcake. How much change do you get?	11p
8	You have 2p, 2p, 2p, 5p. You are given 10p. How much do you have now?	21p
9	Which month comes after August?	September
10	What time is it?	half past two

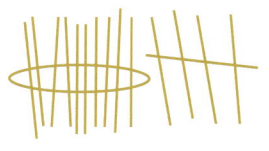

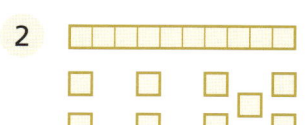

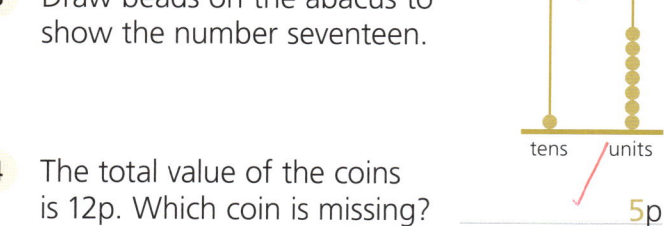

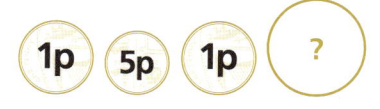

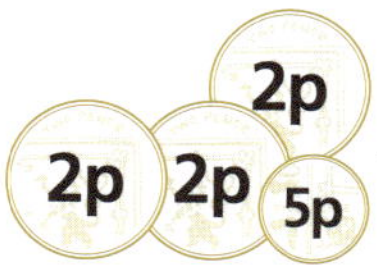

B

#	Question	Answer
1	3 + 10 + 5 =	18
2	23 + 15 =	38
3	7 + 10 + 7 =	24
4	47 + 17 =	64
5	16 − 10 − 3 =	3
6	56 − 13 =	43
7	21 − 10 − 6 =	5
8	31 − 16 =	15
9	☐ − 19 = 4	23
10	☐ + 19 = 24	5

C

#	Question	Answer
1	Add 9 and 14.	23
2	Take 4 from 19.	15
3	The total of 12 and 3 is	15
4	The difference between 18 and 7 is	11
5	The sum of 11 and 2 is	13
6	17 is 8 more than	9
7	Subtract 4 from 13.	9
8	19 minus 6 is	13
9	5 less than 10 is	5
10	15 plus 5 is	20

Mental Arithmetic Introductory Book Answers

SECTION 2 | Test 12

A

#	Question	Answer
1	Turn the picture into a sum.	2 sets of 10
2	Write the number of cubes in words.	forty
3	Draw beads on the abacus to show the number fourteen.	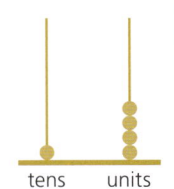
4	The total value of the notes is £60. Which note is missing?	£10
5	How much more do you need to make 20p?	6p
6	You have 2p, 2p, 5p, 10p. You spend 10p. How much have you got left?	9p
7	You have 13p. You buy 8p. How much change do you get?	5p
8	You have 2p, 2p, 2p, 1p, 5p. You win 10p. How much do you have now?	22p
9	Which month comes after February?	March
10	What time is it?	half past six

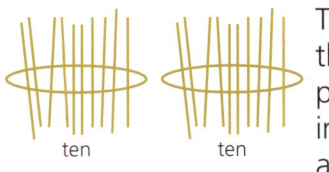

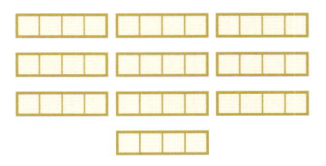

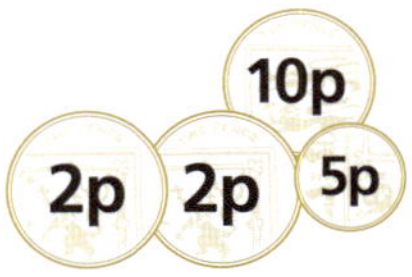

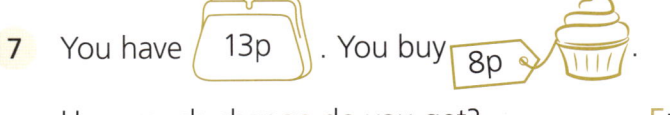

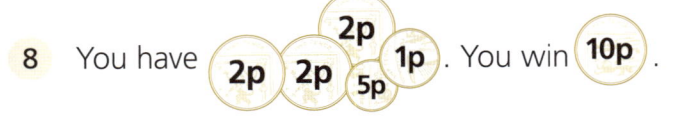

B

#	Question	Answer
1	4 + 10 + 4 =	18
2	44 + 14 =	58
3	10 + 10 + 10 + 10 =	40
4	4 × 10 =	40
5	22 − 10 − 7 =	5
6	32 − 17 =	15
7	40 − 10 − 10 − 10 − 10 =	0
8	40 ÷ 10 =	4
9	☐ − 18 = 6	24
10	☐ + 18 = 26	8

C

#	Question	Answer
1	5 sets of 10 are	50
2	How many tens in 60?	6
3	The total of 5 and 16 is	21
4	The difference between 13 and 7 is	6
5	8 times 10 is	80
6	11 is 8 more than	3
7	Subtract 9 from 14.	5
8	40 divided by 10 is	4
9	3 less than 18 is	15
10	50 shared by 10 is	5

Name of child _____		Diagnostic Chart for Section 2 Indicate where child has difficulty									
		1	2	3	4	5	6	7	8	9	10
Test 1	Part A										
	Part B										
	Part C										
Test 2	Part A										
	Part B										
	Part C										
Test 3	Part A										
	Part B										
	Part C										
Test 4	Part A										
	Part B										
	Part C										
Test 5	Part A										
	Part B										
	Part C										
Test 6	Part A										
	Part B										
	Part C										
Test 7	Part A										
	Part B										
	Part C										
Test 8	Part A										
	Part B										
	Part C										
Test 9	Part A										
	Part B										
	Part C										
Test 10	Part A										
	Part B										
	Part C										
Test 11	Part A										
	Part B										
	Part C										
Test 12	Part A										
	Part B										
	Part C										

From: **Mental Arithmetic Introductory Book Answers**. Copyright © Schofield & Sims Ltd, 2016. This page may be photocopied after purchase.

Mental Arithmetic Introductory Book Answers

These tests enable children to use mental methods to add and subtract multiples of 10.

The tests in Section 3 cover the following topics.

TEST 1	Addition and subtraction of 20	Regard the number as so many sets of 10 and then use single-digit strategies within the tens digits. **Part C** uses language of both measures and number.
TEST 2	Addition and subtraction of 30	
TEST 3	Addition and subtraction of 40	
TEST 4	Addition and subtraction of 50	
TEST 5	Addition and subtraction of 60	
TEST 6	Addition and subtraction of 70	
TEST 7	Addition and subtraction of 80	
TEST 8	Addition and subtraction of 90	
TEST 9	Addition and subtraction of 100	
TEST 10	Sets of 5	Multiplication
TEST 11	A mixture of question types from tests 1 to 9, to test progress	
TEST 12	A mixture of question types from tests 1 to 10, to test progress **Part C of this test is more complicated than the others as it puts together the language learnt, the units used and a possible problem situation.**	

SECTION 3 | Notes on Content

Tests in Section 3

The tests in this section are designed to help children to develop the ability to add and subtract tens and to extend the awareness of place value. It is hoped that these strategies will have been taught previously.

The layout of the tests and the specific focus of each (see contents and the tables below) should help with the diagnosis of particular problems.

Part A
This section uses pictures as follows. Numerical questions relate to the 'ten' number being practised.

Question	
1	identifying the number (words)
2	showing the number on an HTU abacus
3	identifying an abacus number with hundreds
4	making money (from tens number) to £1
5	make number with 10p coins (sets of 10)
6	make number with 5p coins (sets of 5)
7	make number with 2p coins (sets of 2)
8	using 'difference' language with measures
9	finding a number on a number line
10	telling the time to the quarter hour

Part B
This section uses patterns to develop a mental strategy for multiples of ten and extend this to larger numbers.

Question	
1	developing a strategy for addition
2	
3	extending the strategy to larger numbers
4	
5	using addition strategies to complete a sequence
6	developing a strategy for subtraction
7	
8	extending the strategy to larger numbers
9	
10	using subtraction strategies to complete a sequence

Part C
This section uses addition and subtraction language, units of measurement language and measures with multiples of 10. The order is kept the same between tests to make it easy to diagnose consistent mistakes.

Question	
1	add
2	take
3	total
4	difference
5	sum
6	more than
7	subtract
8	minus
9	less than
10	plus

SECTION 3 | Test 1

Schofield & Sims

A | Answer

1. Write the number of sticks in words. __twenty__ (ten, ten)

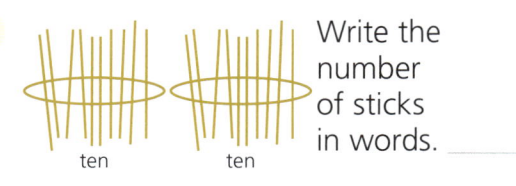

2. Draw beads on the abacus to show the number twenty.

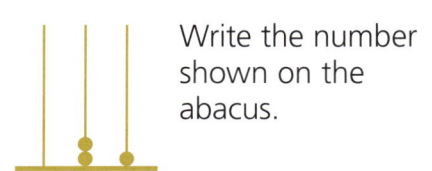

3. Write the number shown on the abacus. __21__

4. How much more do you need to make £1? (20p) __80__p

5. How many 10p make 20p? __2__

6. How many 5p make 20p? __4__

7. How many 2p make 20p? __10__

8. What is the difference between **A** and **B**? __8__cm
   A ———— 20cm
   B ——— 12cm

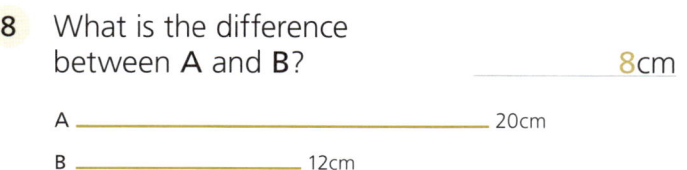

9. What number is the arrow pointing to? __5__
   (number line 0 to 10)

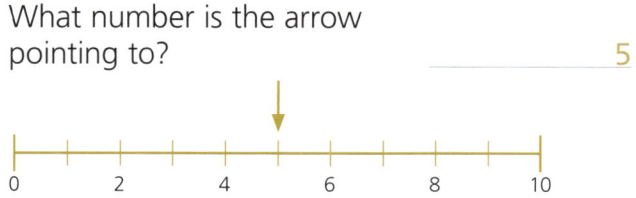

10. 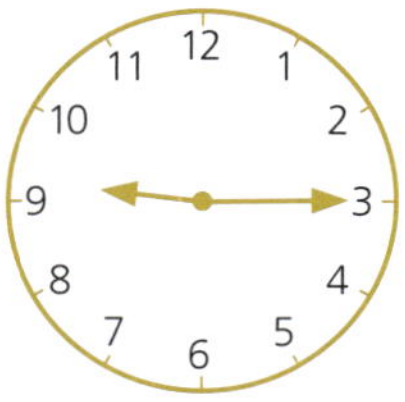 What time is it? __quarter past nine__

B | Answer

1. 7 + 10 + 10 = __27__
2. 17 + 2 tens = __37__
3. 27 + 20 = __47__
4. 37 + 20 = __57__
5. 17, 37, 57, 77, ▢ __97__
6. 24 – 10 – 10 = __4__
7. 24 – 2 tens = __4__
8. 34 – 20 = __14__
9. 44 – 20 = __24__
10. 84, 64, 44, ▢, 4 __24__

C | Answer

1. Add 7p and 20p. __27__p
2. Take 5p from 20p. __15__p
3. The total cost of 20p and 9p is __29__p
4. The difference in price between 7p and 20p is __13__p
5. The sum of 6p and 20p is __26__p
6. 20p is 8p more than __12__p
7. If you spend 10p, your change from 20p is __10__p
8. If you have 20p and you lose 4p, you will be left with __16__p
9. 3p less than 20p is __17__p
10. 12p plus 20p is __32__p

Mental Arithmetic Introductory Book Answers

SECTION 3 | Test 2

A | Answer

1. Write the number of sticks in words. — **thirty**

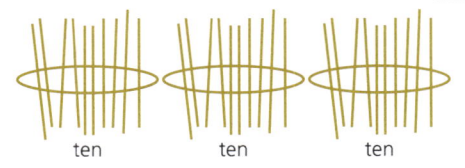

2. Draw beads on the abacus to show the number thirty.

3. 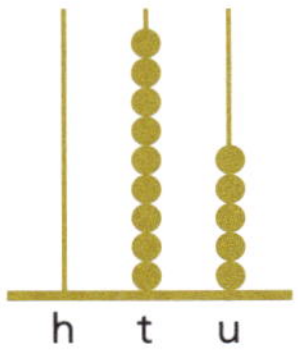 Write the number shown on the abacus. — **95**

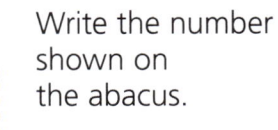

4. How much more do you need to make £1? — **70p**

5. How many 10p make 10p 20p? — **3**

6. How many 5p make 10p 20p? — **6**

7. How many 2p make 10p 20p? — **15**

8. What is the difference in mass between A and B? — **10g**

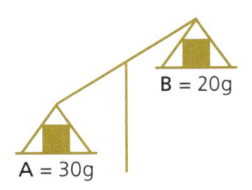

9. What number is the arrow pointing to? — **7**

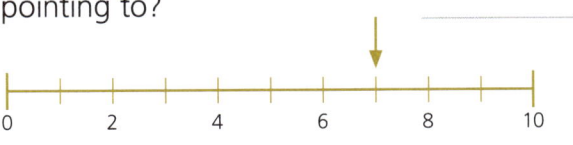

10. 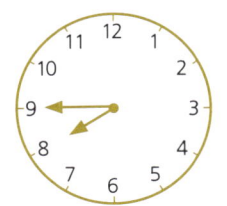 What time is it? — **quarter to eight**

B | Answer

1. 6 + 10 + 10 + 10 = — **36**
2. 16 + 3 tens = — **46**
3. 26 + 30 = — **56**
4. 36 + 30 = — **66**
5. 6, 36, 66, ▢, 126 — **96**
6. 32 − 10 − 10 − 10 = — **2**
7. 42 − 3 tens = — **12**
8. 52 − 30 = — **22**
9. 62 − 30 = — **32**
10. 122, 92, 62, 32, ▢ — **2**

C | Answer

1. Add 7cm and 30cm. — **37cm**
2. Take 5cm from 30cm. — **25cm**
3. The total length of 30cm and 19cm is — **49cm**
4. The difference between 7cm and 30cm is — **23cm**
5. The sum of 16cm and 30cm is — **46cm**
6. 30cm is 8cm longer than — **22cm**
7. Subtract 11cm from 30cm. — **19cm**
8. 30cm minus 14cm is — **16cm**
9. 13cm shorter than 30cm is — **17cm**
10. 20cm plus 30cm is — **50cm**

SECTION 3 | Test 3

Schofield & Sims

A

		Answer
1	Write the number of sticks in words.	forty
2	Draw beads on the abacus to show the number forty.	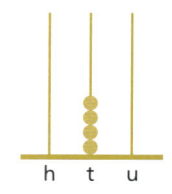
3	Write the number shown on the abacus.	97
4	How much more do you need to make £1? (20p 20p)	60p
5	How many 10p make 20p 20p?	4
6	How many 5p make 20p 20p?	8
7	How many 2p make 20p 20p?	20
8	What is the difference between 20p 20p and 10p?	30p
9	What number is the arrow pointing to?	3
10	What time is it?	quarter past three

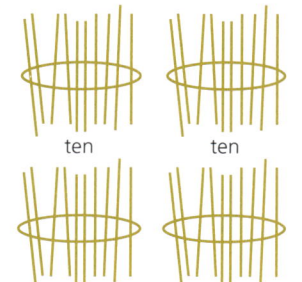

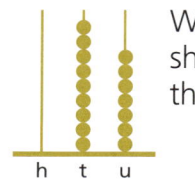

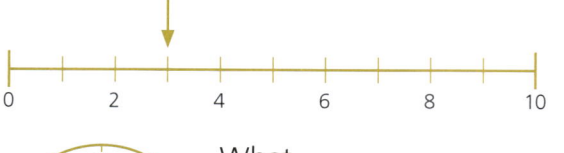

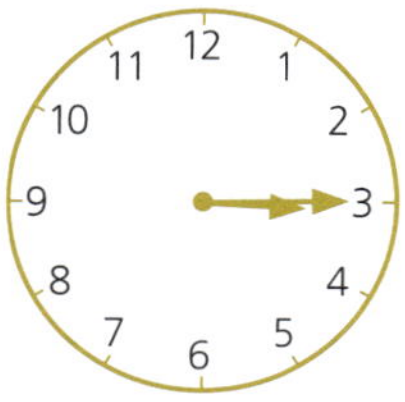

B

		Answer
1	5 + 10 + 10 + 10 + 10 =	45
2	15 + 4 tens =	55
3	25 + 40 =	65
4	35 + 40 =	75
5	5, 25, 45, 65,	85
6	41 − 10 − 10 − 10 − 10 =	1
7	51 − 4 tens =	11
8	61 − 40 =	21
9	71 − 40 =	31
10	101, 81, 61, 41,	21

C

		Answer
1	Add 7kg and 40kg.	47kg
2	Take 5kg from 40kg.	35kg
3	The total mass of 40kg and 20kg is	60kg
4	The difference between 30kg and 40kg is	10kg
5	The sum of 16kg and 40kg is	56kg
6	40kg is 8kg heavier than	32kg
7	Subtract 11kg from 40kg.	29kg
8	40kg minus 4kg is	36kg
9	10kg lighter than 40kg is	30kg
10	12kg plus 40kg is	52kg

Mental Arithmetic Introductory Book Answers

SECTION 3 | Test 4

A — Answer

1. Write the number of cubes in words. — **fifty**
2. Draw beads on the abacus to show the number fifty. 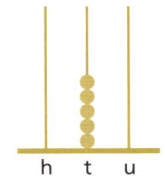
3.  Write the number shown on the abacus. — **99**
4. How much more do you need to make £1? — **50p**
5. How many 10p make 50p? — **5**
6. How many 5p make 50p? — **10**
7. How many 2p make 50p? — **25**
8. What is the difference between A and B? — **50m**
   A ——— 100m
   B ——— 50m 
9. What number is the arrow pointing to? — **9** 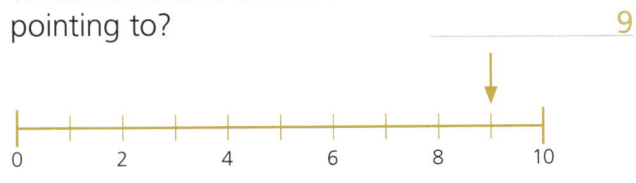
10. 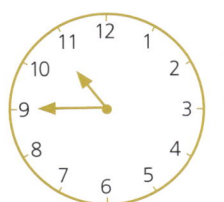 What time is it? — **quarter to eleven**

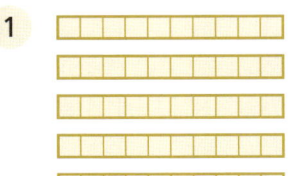

B — Answer

1. 5 + 5 tens = — **55**
2. 15 + 5 tens = — **65**
3. 25 + 50 = — **75**
4. 35 + 50 = — **85**
5. 1, 21, 41, 61, ▢ — **81**
6. 59 − 5 tens = — **9**
7. 69 − 5 tens = — **19**
8. 79 − 50 = — **29**
9. 89 − 50 = — **39**
10. 85, 75, 65, 55, ▢ — **45**

C — Answer

1. Add £30 and £50. — **£80**
2. Take £5 from £50. — **£45**
3. The total value of £50 and £40 is — **£90**
4. The difference between £20 and £50 is — **£30**
5. The sum of £16 and £50 is — **£66**
6. £50 is £8 more than — **£42**
7. If you spend £15, what is your change from £50? — **£35**
8. £50 minus £40 is — **£10**
9. You have £50. You spend £10. How much have you left? — **£40**
10. £19 plus £50 is — **£69**

35

SECTION 3 | Test 5

A

		Answer
1	Write the number of cubes in words.	sixty
2	Draw beads on the abacus to show the number sixty.	
3	Write the number shown on the abacus.	100
4	How much more do you need to make £1?	40p
5	How many 10p make 20p 20p 20p?	6
6	How many 5p make 20p 20p 20p?	12
7	How many 2p make 20p 20p 20p?	30
8	What is the difference in mass between A and B? (A = 60kg, B = 20kg)	40kg
9	What number is the arrow pointing to?	4
10	What time is it?	quarter to twelve

B

		Answer
1	8 + 6 tens =	68
2	18 + 6 tens =	78
3	28 + 60 =	88
4	38 + 60 =	98
5	8, 28, 48, 68,	88
6	60 − 6 tens =	0
7	70 − 6 tens =	10
8	80 − 60 =	20
9	90 − 60 =	30
10	100, 80, 60, 40,	20

C

		Answer
1	Add 40g and 60g.	100g
2	Take 50g from 60g.	10g
3	The total value of 60g and 20g is	80g
4	The difference between 30g and 60g is	30g
5	The sum of 18g and 60g is	78g
6	60g is 8g heavier than	52g
7	Subtract 11g from 60g.	49g
8	60g minus 40g is	20g
9	20g lighter than 60g is	40g
10	30g plus 60g is	90g

Mental Arithmetic Introductory Book Answers

SECTION 3 | Test 6

A | Answer

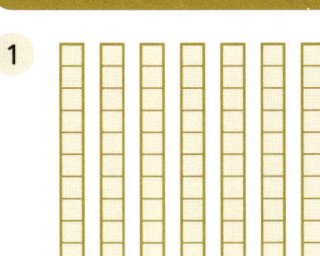

1. Write the number of cubes in words. **seventy**

2. Draw beads on the abacus to show the number seventy.

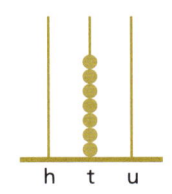

3. Write the number shown on the abacus. **101**

4. 50p 20p — How much more do you need to make £1? **30p**

5. How many 10p make 50p 20p? **7**

6. How many 5p make 50p 20p? **14**

7. How many 2p make 50p 20p? **35**

8. What is the difference between 50p 20p and 10p 10p 10p? **40p**

9. What number is the arrow pointing to? **6**

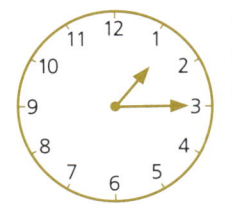

10. What time is it? **quarter past one**

B | Answer

1. 10 + 7 tens = **80**
2. 20 + 7 tens = **90**
3. 30 + 70 = **100**
4. 40 + 70 = **110**
5. 10, 30, 50, 70, **90**
6. 80 – 7 tens = **10**
7. 90 – 7 tens = **20**
8. 100 – 70 = **30**
9. 110 – 70 = **40**
10. 100, 90, 80, 70, **60**

C | Answer

1. Add 30m and 70m. **100m**
2. Take 40m from 70m. **30m**
3. The total length of 70m and 20m is **90m**
4. The difference between 7m and 70m is **63m**
5. The sum of 16m and 70m is **86m**
6. 70m is 20m longer than **50m**
7. Subtract 50m from 70m. **20m**
8. 70m minus 60m is **10m**
9. 8m shorter than 70m is **62m**
10. 19m plus 70m is **89m**

37

SECTION 3 | Test 7

A

1. Write the number of cubes in words. **eighty**
2. Draw beads on the abacus to show the number eighty. 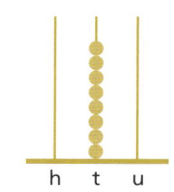
3. 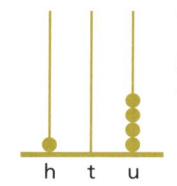 Write the number shown on the abacus. **104**
4. How much more do you need to make £1? **20**p
5. How many  10p make  50p 10p 20p? **8**
6. How many  5p make  20p 20p 20p 20p? **16**
7. How many  2p make  20p 20p 20p 20p? **40**
8. What is the difference between and ? **30** litres
9. What number is the arrow pointing to? **8** 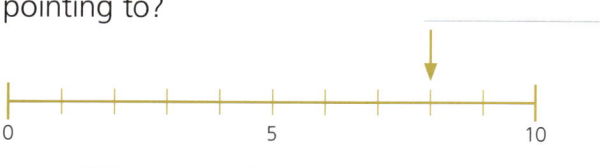
10. 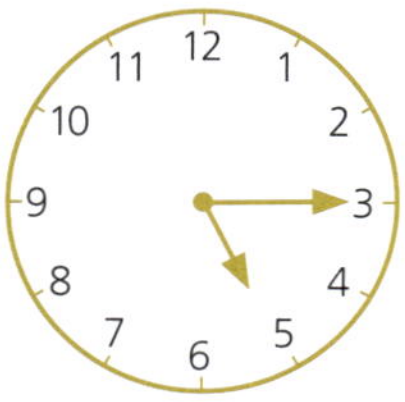 What time is it? **quarter past five**

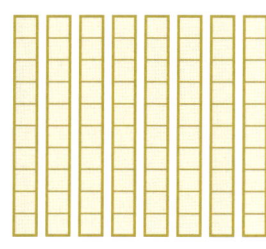

B

1. 10 + 8 tens = **90**
2. 20 + 8 tens = **100**
3. 30 + 80 = **110**
4. 40 + 80 = **120**
5. 0, 30, 60, 90, **120**
6. 80 − 8 tens = **0**
7. 90 − 8 tens = **10**
8. 100 − 80 = **20**
9. 110 − 80 = **30**
10. 101, 91, 81, 71, **61**

C

1. Add 19m and 80m. **99**m
2. Take 60m from 80m. **20**m
3. The total length of 80m and 15m is **95**m
4. The difference between 70m and 80m is **10**m
5. The sum of 6m and 80m is **86**m
6. 80m is 30m longer than **50**m
7. Subtract 20m from 80m. **60**m
8. 80m minus 40m is **40**m
9. 30m shorter than 80m is **50**m
10. 8m plus 80m is **88**m

Mental Arithmetic Introductory Book Answers

SECTION 3 | Test 8

A

1. Write the number of cubes in words. **ninety**
2. Draw beads on the abacus to show the number ninety.
3. Write the number shown on the abacus. **107**
4. How much more do you need to make £1? **10**p
5. How many 10p make 50p 20p 20p? **9**
6. How many 5p make 50p 20p 20p? **18**
7. How many 2p make 50p 20p 20p? **45**
8. What is the difference in mass between A and B? **70**g
9. What number is the arrow pointing to? **3**
10. What time is it? **quarter past seven**

B

1. 10 + 9 tens = **100**
2. 20 + 9 tens = **110**
3. 30 + 90 = **120**
4. 40 + 90 = **130**
5. 0, 25, 50, 75, **100**
6. 100 − 9 tens = **10**
7. 110 − 9 tens = **20**
8. 120 − 90 = **30**
9. 130 − 90 = **40**
10. 105, 85, 65, 45, **25**

C

1. Add 9min and 90min. **99**min
2. Take 50min from 90min. **40**min
3. The total time taken for two TV programmes lasting 90min and 20min is **110**min
4. The difference between 60min and 90min is **30**min
5. The sum of 6min and 90min is **96**min
6. 90min is 5min longer than **85**min
7. Subtract 15min from 90min. **75**min
8. 90min minus 40min is **50**min
9. 30min less than 90min is **60**min
10. 8min plus 90min is **98**min

SECTION 3 | Test 9

Schofield & Sims

A | Answer

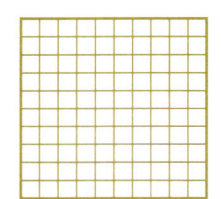

1. Write the number of cubes in words. __one hundred__

2. Draw beads on the abacus to show the number one hundred.

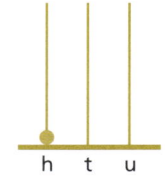

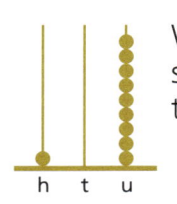

3. Write the number shown on the abacus. __109__

4. How much more do you need to make £1? __45__p

5. How many 10p make £1? __10__

6. How many 5p make £1? __20__

7. How many 2p make £1? __50__

8. What is the difference between [50ml] and [100ml]? __50__ml

9. What number is the arrow pointing to? __1__

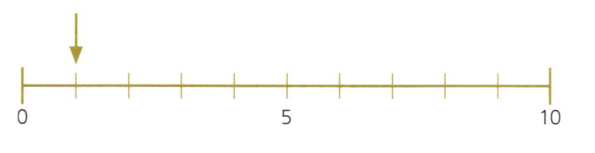

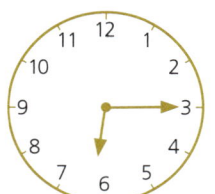

10. What time is it? __quarter past six__

B | Answer

1. 10 + 10 tens = __110__
2. 20 + 1 hundred = __120__
3. 30 + 100 = __130__
4. 40 + 100 = __140__
5. 0, 100, 200, 300, __400__
6. 100 – 10 tens = __0__
7. 110 – 1 hundred = __10__
8. 120 – 100 = __20__
9. 130 – 100 = __30__
10. 900, 800, 700, 600, __500__

C | Answer

1. Add 7 tens and 10 tens. __170__
2. Take 5 tens from 10 tens. __50__
3. The total of 10 tens and 9 tens is __190__
4. The difference between 7 tens and 10 tens is __30__
5. The sum of 6 tens and 10 tens is __160__
6. 10 tens are 8 tens more than __20__
7. Subtract 1 ten from 10 tens. __90__
8. 10 tens minus 4 tens is __60__
9. 3 tens fewer than 10 tens is __70__
10. 2 tens plus 10 tens is __120__

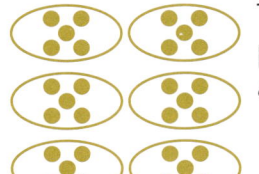

Mental Arithmetic Introductory Book Answers

SECTION 3 | Test 10

A

1. Turn the picture into a sum. 6 sets of 5
2. Turn the picture into a sum. 5 sets of 6
3. × 5 = 45
4. ÷ 5 = 3
5. How much altogether? 25p 
6. 5p × ☐ = 40p 8
7. 20p ÷ 5 = 4p
8. B is five times as heavy as A. How heavy is B? (A = 9g) 45g
9. What number is the arrow pointing to? 2 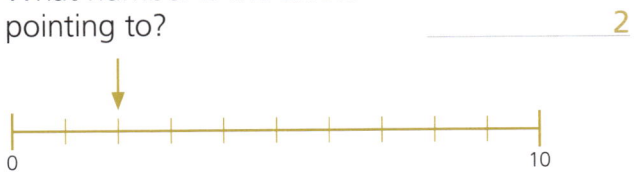
10. 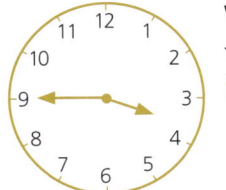 What time is it? quarter to four

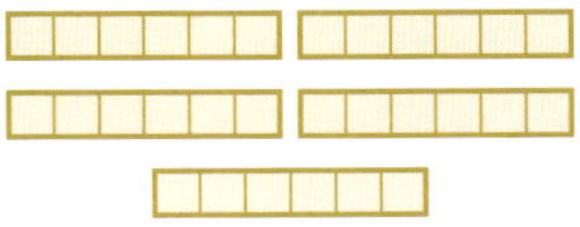

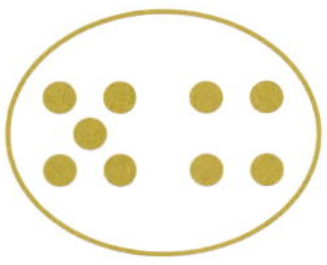

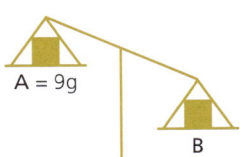

B

1. 5, 10, 15, 20, ☐ 25
2. 5 + 5 + 5 + 5 = 20
3. 4 × 5 = 20
4. 5 + 5 + 5 + 5 + 5 + 5 = 30
5. 6 × 5 = 30
6. 50, 45, 40, 35, ☐ 30
7. 15 − 5 − 5 − 5 = 0
8. 15 ÷ 5 = 3
9. 35 ÷ 5 = 7
10. 40 ÷ 5 = 8

C

1. 8 sets of 5 are 40
2. How many fives in 30? 6
3. 9 fives are 45
4. How many 5p coins make 50p? 10
5. 7 times 5 is 35
6. 6 multiplied by 5 is 30
7. 40 shared among 5 is 8
8. If I change 20p into 5p coins, how many will I get? 4
9. 45 divided by 5 is 9
10. How many toes are there on 10 feet? 50

41

SECTION 3 | Test 11

A

1. Write the number of cubes in words. — **thirty**

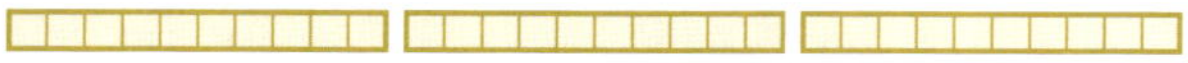

2. Draw beads on the abacus to show the number sixty-four.

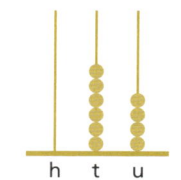

3. Write the number shown on the abacus. — **110**

4. 50p 10p 10p — How much more do you need to make £1? — **30**p

5. How many 10p make 20p 20p 20p 20p? — **8**

6. How many 5p make 10p 10p? — **4**

7. How many 2p make 10p 10p 20p? — **20**

8. What is the difference between £1 and 50p? — **50**p

9. What number is the arrow pointing to? — **4**

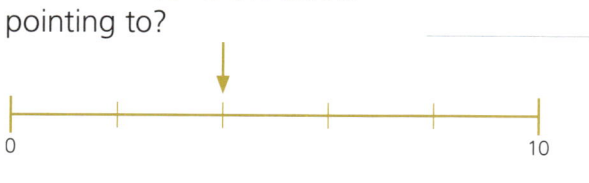

10. What time is it? — **quarter past two**

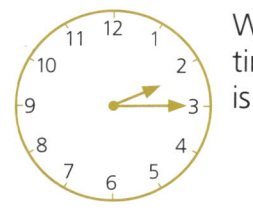

B

1. 5 + 5 tens = — **55**
2. 35 + 50 = — **85**
3. 3 + 7 tens = — **73**
4. 23 + 70 = — **93**
5. 3, 23, 43, 63, ☐ — **83**
6. 78 − 3 tens = — **48**
7. 98 − 30 = — **68**
8. 90 − 4 tens = — **50**
9. 60 − 40 = — **20**
10. 87, 77, 67, 57, ☐ — **47**

C

1. Add 50cm and 60cm. — **110**cm
2. Take 70 litres from 80 litres. — **10** litres
3. The total mass of four boxes each weighing 10kg is — **40**kg
4. The difference in value between 80p and 30p is — **50**p
5. The sum of 100g and 90g is — **190**g
6. 100 is 5 tens more than — **50**
7. Subtract 30ml from 40ml. — **10**ml
8. 80min minus 50min is — **30**min
9. 70m is 10m shorter than — **80**m
10. £60 plus £20 is — £**80**

42

Mental Arithmetic Introductory Book Answers

SECTION 3 | Test 12

A | Answer

1. Write the number of cubes in words. — forty-two
2. Draw beads on the abacus to show the number thirty-five.
3. Write the number shown on the abacus. — 111
4. How much more do you need to make £1? (20p, 5p) — 75p
5. How many 10p make 50p 10p 10p? — 7
6. How many 5p make 5p 20p? — 5
7. How many 2p make 10p 20p 20p? — 25
8. What is the difference in width between A and B? (width 40cm, width 90cm) — 50cm
9. What number is the arrow pointing to? — 8
10. What time is it? — quarter to nine

B | Answer

1. 4 + 8 tens = — 84
2. 24 + 80 = — 104
3. 6 + 2 tens = — 26
4. 76 + 20 = — 96
5. 5, 105, 205, 305, — 405
6. 87 − 6 tens = — 27
7. 67 − 60 = — 7
8. 99 − 3 tens = — 69
9. 69 − 3 tens = — 39
10. 200, 150, 100, — 50

C | Answer

1. On Tuesday a farmer adds 40 litres of milk to 10 litres he got on Monday. How much has he got now? — 50 litres
2. If I take £20 from my savings of £80, how much is left? — £60
3. The total length of two sticks measuring 70cm and 60cm is — 130cm
4. The difference in price between two comics costing 90p and 30p is — 60p
5. The sum of two loads with masses 50kg and 60kg is — 110kg
6. A 100ml bottle of water has 10ml more than one which contains — 90ml
7. If I cut off 30m from a 90m length of rope, how much is left? — 60m
8. If 7g of a 100g packet of sweets is the wrapper, how much do the sweets weigh? — 93g
9. I scored 2 tens less than Leo who scored 5 tens. What was my score? — 30
10. The time taken for a football match was 40min plus 10min half time. How long did it take altogether? — 50min

43

Name of child _____		Diagnostic Chart for Section 3 Indicate where child has difficulty									
		1	2	3	4	5	6	7	8	9	10
Test 1	Part A										
	Part B										
	Part C										
Test 2	Part A										
	Part B										
	Part C										
Test 3	Part A										
	Part B										
	Part C										
Test 4	Part A										
	Part B										
	Part C										
Test 5	Part A										
	Part B										
	Part C										
Test 6	Part A										
	Part B										
	Part C										
Test 7	Part A										
	Part B										
	Part C										
Test 8	Part A										
	Part B										
	Part C										
Test 9	Part A										
	Part B										
	Part C										
Test 10	Part A										
	Part B										
	Part C										
Test 11	Part A										
	Part B										
	Part C										
Test 12	Part A										
	Part B										
	Part C										

From: **Mental Arithmetic Introductory Book Answers**. Copyright © Schofield & Sims Ltd, 2016. This page may be photocopied after purchase.

Mental Arithmetic Introductory Book Answers

JUST FACTS

DOUBLES

1 + 1 =	2	2 − 1 =	1
2 + 2 =	4	4 − 2 =	2
3 + 3 =	6	6 − 3 =	3
4 + 4 =	8	8 − 4 =	4
5 + 5 =	10	10 − 5 =	5
6 + 6 =	12	12 − 6 =	6
7 + 7 =	14	14 − 7 =	7
8 + 8 =	16	16 − 8 =	8
9 + 9 =	18	18 − 9 =	9
10 + 10 =	20	20 − 10 =	10

NEAR DOUBLES

1 + 2 =	3	2 + 1 =	3	3 − 1 =	2	3 − 2 =	1
2 + 3 =	5	3 + 2 =	5	5 − 2 =	3	5 − 3 =	2
3 + 4 =	7	4 + 3 =	7	7 − 3 =	4	7 − 4 =	3
4 + 5 =	9	5 + 4 =	9	9 − 4 =	5	9 − 5 =	4
5 + 6 =	11	6 + 5 =	11	11 − 5 =	6	11 − 6 =	5
6 + 7 =	13	7 + 6 =	13	13 − 6 =	7	13 − 7 =	6
7 + 8 =	15	8 + 7 =	15	15 − 7 =	8	15 − 8 =	7
8 + 9 =	17	9 + 8 =	17	17 − 8 =	9	17 − 9 =	8

NEAR NEAR DOUBLES

1 + 3 =	4	3 + 1 =	4	4 − 1 =	3	4 − 3 =	1
2 + 4 =	6	4 + 2 =	6	6 − 2 =	4	6 − 4 =	2
3 + 5 =	8	5 + 3 =	8	8 − 3 =	5	8 − 5 =	3
4 + 6 =	10	6 + 4 =	10	10 − 4 =	6	10 − 6 =	4
5 + 7 =	12	7 + 5 =	12	12 − 5 =	7	12 − 7 =	5
6 + 8 =	14	8 + 6 =	14	14 − 6 =	8	14 − 8 =	6
7 + 9 =	16	9 + 7 =	16	16 − 7 =	9	16 − 9 =	7
8 + 10 =	18	10 + 8 =	18	18 − 8 =	10	18 − 10 =	8

JUST FACTS

Schofield & Sims

COUNTING ON 1 and related facts

1 + 1 =	2	1 + 1 =	2	2 − 1 =	1	2 − 1 =	1
2 + 1 =	3	1 + 2 =	3	3 − 1 =	2	3 − 2 =	1
3 + 1 =	4	1 + 3 =	4	4 − 1 =	3	4 − 3 =	1
4 + 1 =	5	1 + 4 =	5	5 − 1 =	4	5 − 4 =	1
5 + 1 =	6	1 + 5 =	6	6 − 1 =	5	6 − 5 =	1
6 + 1 =	7	1 + 6 =	7	7 − 1 =	6	7 − 6 =	1
7 + 1 =	8	1 + 7 =	8	8 − 1 =	7	8 − 7 =	1
8 + 1 =	9	1 + 8 =	9	9 − 1 =	8	9 − 8 =	1
9 + 1 =	10	1 + 9 =	10	10 − 1 =	9	10 − 9 =	1

COUNTING ON 2 and related facts

1 + 2 =	3	2 + 1 =	3	3 − 2 =	1	3 − 1 =	2
2 + 2 =	4	2 + 2 =	4	4 − 2 =	2	4 − 2 =	2
3 + 2 =	5	2 + 3 =	5	5 − 2 =	3	5 − 3 =	2
4 + 2 =	6	2 + 4 =	6	6 − 2 =	4	6 − 4 =	2
5 + 2 =	7	2 + 5 =	7	7 − 2 =	5	7 − 5 =	2
6 + 2 =	8	2 + 6 =	8	8 − 2 =	6	8 − 6 =	2
7 + 2 =	9	2 + 7 =	9	9 − 2 =	7	9 − 7 =	2
8 + 2 =	10	2 + 8 =	10	10 − 2 =	8	10 − 8 =	2
9 + 2 =	11	2 + 9 =	11	11 − 2 =	9	11 − 9 =	2

ADDING 10 and related facts

1 + 10 =	11	10 + 1 =	11	11 − 10 =	1	11 − 1 =	10
2 + 10 =	12	10 + 2 =	12	12 − 10 =	2	12 − 2 =	10
3 + 10 =	13	10 + 3 =	13	13 − 10 =	3	13 − 3 =	10
4 + 10 =	14	10 + 4 =	14	14 − 10 =	4	14 − 4 =	10
5 + 10 =	15	10 + 5 =	15	15 − 10 =	5	15 − 5 =	10
6 + 10 =	16	10 + 6 =	16	16 − 10 =	6	16 − 6 =	10
7 + 10 =	17	10 + 7 =	17	17 − 10 =	7	17 − 7 =	10
8 + 10 =	18	10 + 8 =	18	18 − 10 =	8	18 − 8 =	10
9 + 10 =	19	10 + 9 =	19	19 − 10 =	9	19 − 9 =	10

Mental Arithmetic Introductory Book Answers

JUST FACTS

MAKING 10 and related facts

1 + 9 =	10	9 + 1 =	10	10 − 1 =	9	10 − 9 =	1
2 + 8 =	10	8 + 2 =	10	10 − 2 =	8	10 − 8 =	2
3 + 7 =	10	7 + 3 =	10	10 − 3 =	7	10 − 7 =	3
4 + 6 =	10	6 + 4 =	10	10 − 4 =	6	10 − 6 =	4
5 + 5 =	10	5 + 5 =	10	10 − 5 =	5	10 − 5 =	5

ADDING 9 and related facts

1 + 9 =	10	9 + 1 =	10	10 − 9 =	1	10 − 1 =	9
2 + 9 =	11	9 + 2 =	11	11 − 9 =	2	11 − 2 =	9
3 + 9 =	12	9 + 3 =	12	12 − 9 =	3	12 − 3 =	9
4 + 9 =	13	9 + 4 =	13	13 − 9 =	4	13 − 4 =	9
5 + 9 =	14	9 + 5 =	14	14 − 9 =	5	14 − 5 =	9
6 + 9 =	15	9 + 6 =	15	15 − 9 =	6	15 − 6 =	9
7 + 9 =	16	9 + 7 =	16	16 − 9 =	7	16 − 7 =	9
8 + 9 =	17	9 + 8 =	17	17 − 9 =	8	17 − 8 =	9
9 + 9 =	18	9 + 9 =	18	18 − 9 =	9	18 − 9 =	9

ADDING 8 and related facts

1 + 8 =	9	8 + 1 =	9	9 − 8 =	1	9 − 1 =	8
2 + 8 =	10	8 + 2 =	10	10 − 8 =	2	10 − 2 =	8
3 + 8 =	11	8 + 3 =	11	11 − 8 =	3	11 − 3 =	8
4 + 8 =	12	8 + 4 =	12	12 − 8 =	4	12 − 4 =	8
5 + 8 =	13	8 + 5 =	13	13 − 8 =	5	13 − 5 =	8
6 + 8 =	14	8 + 6 =	14	14 − 8 =	6	14 − 6 =	8
7 + 8 =	15	8 + 7 =	15	15 − 8 =	7	15 − 7 =	8
8 + 8 =	16	8 + 8 =	16	16 − 8 =	8	16 − 8 =	8
9 + 8 =	17	8 + 9 =	17	17 − 8 =	9	17 − 9 =	8

Full list of Schofield & Sims Mental Arithmetic books

Pupil books

Mental Arithmetic Introductory Book	ISBN 978 07217 0798 3
Mental Arithmetic 1	ISBN 978 07217 0799 0
Mental Arithmetic 2	ISBN 978 07217 0800 3
Mental Arithmetic 3	ISBN 978 07217 0801 0
Mental Arithmetic 4	ISBN 978 07217 0802 7
Mental Arithmetic 5	ISBN 978 07217 0803 4
Mental Arithmetic 6	ISBN 978 07217 0804 1

Answer books

Mental Arithmetic Introductory Book Answers	ISBN 978 07217 0853 9
Mental Arithmetic 1 Answers	ISBN 978 07217 0805 8
Mental Arithmetic 2 Answers	ISBN 978 07217 0806 5
Mental Arithmetic 3 Answers	ISBN 978 07217 0807 2
Mental Arithmetic 4 Answers	ISBN 978 07217 0808 9
Mental Arithmetic 5 Answers	ISBN 978 07217 0809 6
Mental Arithmetic 6 Answers	ISBN 978 07217 0810 2

Teacher's Guide

Mental Arithmetic Teacher's Guide	ISBN 978 07217 1389 2

Free downloads

A range of free downloads is available from the Schofield & Sims website (www.schofieldandsims.co.uk). These downloads may be used to support pupils in their learning, both in school and at home. They include the following items:

- two **Mental Arithmetic** Entry Tests to help you choose the best book for each individual
- an Achievement Award certificate for each **Mental Arithmetic** book
- Maths Facts downloads to provide a quick reference tool
- a National Curriculum Chart to show how each book supports the programmes of study.